TRANSPORTATION

and

TRAFFIC MANAGEMENT

by

E. ALBERT OVENS

VOLUME THREE

1980

Tenth Edition

Educational Division of The Traffic Service Corporation

College of Advanced Traffic
22 West Madison St.
Chicago, Illinois 60602
Telephone (312) 346-8630

Academy of Advanced Traffic
One World Trade Center, Suite 5457
New York, NY 10048
Telephone (212) 466-1980

This volume, together with Volumes I, II, and IV, comprise the theory portion of the 4-semester course in Transportation and Traffic Management offered by the College of Advanced Traffic and the Academy of Advanced Traffic.

Library of Congress Card No. 74-19874

ISBN 0-87408-012-6

The editor of this book was Donald V. Keyes.
Produced by Stephen R. Hunter.
It was set in Times Roman by the Traffic Service Corporation.

CONTENTS

CHAPTER 1

OCEAN CARRIER TARIFFS AND RATES

General

The interstate freight rates of the railroads are stable throughout the entire country and may not be changed except upon thirty days' notice, unless shorter notice is authorized by the Interstate Commerce Commission. This is true of all other carriers regulated by the Commission, including inland, coastwise and intercoastal water carriers. The rates of ocean carriers, on the other hand, are subject to greater fluctuation. Thirty days' notice must be given in the case of rate increases, but reductions can be made on one day's notice to the Federal Maritime Commission. Under these conditions, the demand for cargo space effectively controls the level of rates. If there is little demand for space, rates can be adjusted downward very quickly, and while the reduced rates have to be maintained for thirty days, they can be increased after that time if demand conditions permit. Ocean carriers have attempted to stabilize conditions as much as possible through working arrangements with other lines in so-called conferences. These conferences include both United States and foreign lines.

In foreign trade one will work with metric weights and measures and cubic feet of space for cargo, as well as our usual American system of pounds and cubic feet. Frequently, when a steamship company is asked for a quotation of freight rates it names a rate "per ton" weight or measurements, "ship's option." This means that the freight usually will be assessed either per ton of 2000 or 2240 pounds or per measurement ton of 40 cubic feet, whichever may be found to be more advantageous for the ship, that is, bring in the greater revenue. If, in steamship parlance, a package "measures more than it weighs," that is, if its weight for example is only 36 pounds to the cubic foot, then the freight rate will be per ton of 40 cubic feet. Since comparatively few commodities weight more than they measure, it follows that in the great majority of cases ocean freight charges are based on a measurement ton rate and not on a weight ton rate.

It is quite likely, however, that with the growth of containerization and the establishment of through intermodal rates the ship's option of weight or measurement basis of ocean freight rates will be encountered only in connection with break-bulk movements.

Jurisdiction of Federal Maritime Commission

The Merchant Marine Shipping Act, 1936, created the United States Maritime Commission, now known as the Federal Maritime Commission, and transferred to this Commission all the functions, powers, and duties vested in the former United States Shipping Board by the Shipping Act, 1916, the Merchant Marine Act, 1920, the Merchant Marine Act, 1928, the International Shipping Act, 1933, and amendments to those acts. This Act also, under Section 205, made it unlawful for any common carrier by water, either directly or indirectly, through the medium of an agreement, conference, association, understanding or otherwise, to prevent or attempt to prevent any other such carrier from serving any port designed for the accommodation of ocean-going vessels located on any improvement project by the Congress or through it by any other agency of the Federal Government, lying within the continental limits of the United States, at the same rates which it charges at the nearest port already regularly served by it.

Under the Shipping Act of 1916, as amended, the regulatory powers of the Federal Maritime Commission apply to ocean carriers engaged in both domestic offshore commerce and foreign commerce. The domestic offshore commerce includes operations between points in continental United States and points in such non-contiguous states, territories, and possessions as Hawaii, Alaska, Puerto Rico, Guam, Wake and the Virgin Islands. The F.M.C. jurisdiction also extends to ocean freight forwarders (referred to as "non-vessel operating common carriers by water," NVO's or NVOCC's). These ocean freight forwarders are not subject to any certification requirements. They hold themselves out to provide transportation under a tariff filed with F.M.C. without acting as the underlying carrier. Their operating margins are the difference between the rates they charge their shippers and the rates charged them by the underlying carrier.

As far as rates and tariffs are concerned the ocean carriers are required (Section 18) to publish tariffs and to file them with the F.M.C. They are not to deviate from the rates and practices shown in their tariffs and the F.M.C. has the authority to disapprove and set maximum rates, fares and charges in domestic commerce. They can also disapprove a rate applying to foreign commerce as being either so unreasonably high or low as to be detrimental to the commerce of the United States. New rates and increased rates in foreign commerce may not become effective until thirty days after tariff publication, except upon permission of the F.M.C. but decreased rates can become effective immediately.

Section 18 of the Shipping Act also authorizes the F.M.C. to prescribe regulations governing the form and manner in which tariffs shall be published and filed. These regulations are published in Part 536 of Title 46, Code of Federal Regulations, and are similar, in many respects, to those of the Interstate Commerce Commission.

Sections 14, 14a and 14b of the Shipping Act are worthy of special attention as they are uniquely different from anything we have encountered thus far in our study of inland transportation and the regulation thereof. The full text of these sections is as follows:

> "Sec. 14. That no common carrier by water shall, directly or indirectly, in respect to the transportation by water of passengers or property between a port of a State, Territory, District, or possession of the United States and any other such port or a port of a foreign country—
>
> "First. Pay, or allow, or enter into any combination, agreement, or understanding, express or implied, to pay or allow, a deferred rebate to any shipper. The term 'deferred rebate' in this Act means a return of any portion of the freight money by a carrier to any shipper as a consideration for the giving of all or any portion of his shipments to the same or any other carrier, or for any other purpose, the payment of which is deferred beyond the completion of the service for which it is paid, and is made only if, during both the period for which computed and the period of deferment, the shipper has complied with the terms of the rebate agreement or arrangement.
>
> "Second. Use a fighting ship either separately or in conjunction with any other carrier, through agreement or otherwise. The term 'fighting ship' in this Act means a vessel used in a particular trade by a carrier or group of carriers for the purpose of excluding, preventing, or reducing competition by driving another carrier out of such trade.
>
> "Third. Retaliate against any shipper by refusing or threatening to refuse space accommodations when such are available, or resort to other discriminating or unfair methods, because such shipper has patronized any other carrier or has filed a complaint charging unfair treatment, or for any other reason.
>
> "Fourth. Make any unfair or unjustly discriminatory contract with any shipper based on the volume of freight offered, or unfairly treat or unjustly discriminate against any shipper in the matter of (a) cargo space accommodations or other facilities, due regard being had for the proper loading of the vessel and the available tonnage; (b) the loading and landing of freight in proper condition; or (c) the adjustment and settlement of claims.
>
> "Any carrier who violates any provision of this section shall be guilty of a misdemeanor punishable by a fine of not more than $25,000 for each of-

fense: Provided, That nothing in this section or elsewhere in this Act shall be construed or applied to forbid or make unlawful any dual rate contract arrangement in use by the members of a conference on May 19, 1958, which conference is organized under an agreement approved under Section 15 of this Act by the regulatory body administering this Act, unless and until such regulatory body disapproves, cancels, or modifies such arrangement in accordance with the standards set forth in Section 15 of this Act. the term 'dual rate contract arrangement' as used herein means a practice whereby a conference establishes tariffs of rates at two levels, the lower of which will be charged to merchants who agree to ship their cargoes on vessels of members of the conference only, and the higher of which shall be charged to merchants who do not so agree.

"Sec. 14a. The board upon its own initiative may, or upon complaint shall, after due notice to all parties in interest and hearing, determine whether any person, not a citizen of the United States and engaged in transportation by water of passengers or property—

"(1) Has violated any provision of Section 14, or

"(2) Is a party to any combination, agreement, or understanding, express or implied, that involves in respect to transportation of passengers or property between foreign ports, deferred rebates or any other unfair practice designated in Section 14, and that excludes from admission upon equal terms with all other parties thereto, a common carrier by water which is a citizen of the United States and which has applied for admission.

"If the board determines that any such person has violated any such provision or is a party to any such combination, agreement, or understanding, the board shall thereupon certify such fact to the Secretary of Commerce. The Secretary shall thereafter refuse such person the right of entry for any ship owned or operated by him or any carrier directly or indirectly controlled by him, into any port of the United States, or any Territory, District, or possession thereof, until the board certifies that the violation has ceased or such combination, agreement, or understanding has been terminated.

"Sec. 14b. Notwithstanding any other provisions of this Act, on application the Federal Maritime Commission (hereinafter 'Commission'), shall, after notice and hearing, by order, permit the use of any common carrier or conference of such carriers in foreign commerce of any contract, amendment, or modification thereof, which is available to all shippers and consignees on equal terms and conditions, which provides lower rates to a shipper or consignee who agrees to give all or any fixed portion of his patronage to such carrier or conference of carriers unless the Commission finds that the contract, amendment, or modification thereof will be detrimental to the commerce of the United States or contrary to the public interest, or unjustly discriminatory or unfair as between shippers, exporters, importers, or ports, or between exporters from the United States and their foreign competitors, and provided the contract, amendment or modification thereof expressly (1) permits prompt release of the contract shipper from the contract with respect to any shipment or shipments for which the contracting carrier or conference of carriers cannot provide as much space as the contract shipper shall require on reasonable notice; (2) provides that whenever a tariff rate for the carriage of goods under the contract becomes effective, insofar as it is under the control of the carrier or conference of carriers, it shall not be increased before a reasonable period, but in no case less than ninety days; (3) covers only those goods of the contract shipper as to the shipment of which he has the legal right at the time of shipment to select the carrier: Provided, however, That it shall be deemed a breach of the contract if, before the time of shipment and with the intent to avoid his obligation under the contract, the contract shipper divests himself, or with the same intent permits himself to be divested, of the legal right to select the carrier and the shipment is carried by a carrier which is not a party to the contract; (4) does not require the contract shipper to divert shipment of goods from natural routings not served by the carrier or conference of carriers where direct carriage is available; (5) limits damages recoverable for breach by either party to actual damages to be determined after breach in accordance with the principles of contract law: Provided, however, That the contract may specify that in the case of a breach by a contract shipper the damages may be an amount not exceeding the freight charges computed at the contract rate on the particular shipment, less the cost of handling; (6) permits the contract shipper to terminate at any time without penalty upon ninety days' notice; (7) provides for a spread between ordinary rates and rates charged contract shippers which the Commission finds to be reasonable in all the circumstances but which spread shall in no event be more than 15 per centum of the ordinary rates; (8) excludes cargo of the contract shippers which is loaded and carried in bulk without mark or count except liquid bulk cargoes, other than chemicals, in less than full shipload lots: Provided, however, That upon finding that economic factors so warrant, the Commission may exclude from the contract any commodity subject to the foregoing exception; and (9) contains such other provisions not inconsistent herewith as the Commission shall require or permit.

"The Commission shall withdraw permission which it has granted under the authority contained in this section for the use of any contract if it finds, after notice and hearing, that the use of such contract is detrimental to the commerce of the United States or contrary to the public interest, or is unjustly discriminatory or unfair as between shippers, exporters, importers, or ports, or between exporters from the United States and their foreign competitors. The carrier or conference of carriers may on ninety days' notice terminate without penalty the contract rate sysem herein authorized, in whole or with respect to any commodity: Provided, however, That after such termination the carrier or conference of carriers may not reinstitute such contract rate system or part thereof so terminated without prior permission by the Commission in accordance with the provisions of this section. Any contract, amendment, or modification of any contract not permitted by the Commission shall be unlawful, and contracts, amendments, and modifications shall be lawful only when and as long as permitted by the Commission; before permission is granted or after permission is withdrawn it shall be unlawful to carry out in whole or in part, directly or indirectly, any such contract, amendment, or modification. As used in this section, the term 'contract shipper' means a person other than a carrier or conference of carriers who is a party to a contract the use of which may be permitted under this section."

Where the laws, regulations, or practices of any foreign Government operate in such measure that vessels of the United States are not accorded equal privileges in foreign trade with vessels of such foreign countries or vessels of other foreign countries, it is the duty of the Maritime Commission to investigate the action and report the results of such investigation to the President with its recommendations, and the President is, under the Act, authorized and empowered to secure by diplomatic action equal privileges for vessels of the United States. If by such diplomatic action the President shall be unable to secure equal privileges then the President shall advise Congress as to the facts and his conclusions in order that proper action may be taken thereon.

Ocean Carrier Tariffs

As far as form is concerned, the ocean carrier tariffs are not very different from those of the rail or motor carriers. One important difference is that the Federal Maritime Commission tariff regulations require that tariffs shall be in loose leaf form and shall be printed on one side only. In many water carrier tariffs only commodity rates are published and where this is the case it is important, in checking a rate, to look for an item naming a rate on "Commodities, N.O.S." or "Freight, All Kinds" in addition to a specific entry for the commodity shipped. The water carriers do not publish a classification similar to the Uniform Freight Classification of the railroads or the National Motor Freight Classification of the motor carriers. Where class rates are published in a water carrier tariff, Part 536.4 of the Code of Federal Regulations states that, "Class rate tariffs and tariffs containing both class and commodity rates shall contain, in addition to the item or page reference where applicable, the ratings of commodities to which class rates apply. Such index may be omitted where rates on less than 25 commodities are included in the tariff." As with rail and motor carrier tariffs, both arbitraries and differentials are employed where necessary. Special rates are also published such as ad valorem rates for commodities the value of which exceeds the bill of lading liability limits. Also, where necessary, special rates are provided for refrigerator service and for the transportation of dangerous articles. In the case of the break-bulk carrier, rates may differ depending on whether the cargo is stowed in the hold of the vessel or on deck.

In addition to the rate sections, the F.M.C. rules require a table of contents and here again we want to emphasize the importance of the table of contents when using a tariff. Also required are an index of commodities for which rates are provided in the tariff and a section of rules governing the services offered by the participating carriers.

The development and growth of container-ship operations has created additional rate and tariff problems. It is now possible for a shipper to load freight in a container at some inland point in the United States, have it move in piggy-back service or by truck to a United States port where, without breaking bulk, the container is loaded on board a container-ship for transportation to a foreign port and thence on to some inland destination. With the increasing use of such through intermodal services it was inevitable that shippers would request the establishment of through rates, at least from the point of origin in the United States to the foreign port. The various carriers were receptive to publication of through rates, however, the question was, with whom should such rates be filed? Should it be the Interstate Commerce Commission, which has jurisdiction over the originating carriers, or the Federal Maritime Commission, with jurisdiction over the ocean carriers. Unfortunately, the tariff construction and filing requirements of the two regulatory agencies differed in certain respects. In July of 1969 the Interstate Commerce Commission instituted an investigation titled, "In the Matter of Tariffs Containing Joint Rates and Through Routes Between Points in the United States and Points in Foreign Countries," Ex Parte 261. The Commission issued several reports in this proceeding, the last of which was dated January 30, 1976, in which they said:

> "As we stated in our prior reports, our jurisdiction to regulate foreign commerce, insofar as such transportation takes place within the United States, is conferred by Sections 1(1), 203(a)(11), and 302(i)(3) of the Interstate Commerce Act. In our consideration of the intermodal tariffs filed with this Commission, however, we do not intend to assert jurisdiction over or otherwise to engage in substantive regulation of the ocean portion of the rates pursuant to the Interstate Commerce Act. To do so, we believe,

would require us to interpret provisions of the acts administered by the F.M.C. which are within the expertise of that agency. Therefore, we wish to emphasize the fact that our jurisdiction will be invoked solely to accomplish substantive regulation of the domestic carrier's portion of the through rate and that any procedural requirements we may impose will be directed to this end."

A somewhat more detailed discussion of the Commission's decision in this proceeding will be found in a subsequent chapter on the subject of through routes.

The rules of these joint intermodal tariffs are in many ways unusual because of the nature of the services being offered by the carriers party to the tariffs and should be reviewed carefully. This is particularly true of the liability provisions for loss and/or damage to the freight. In many instances the terms and conditions of the through bill of lading are reproduced in the rate tariff, and included in these terms and conditions will be a provision such as the following:

"Subject to all rights, privileges and limitations of and exonerations from liability granted to the carrier under this bill of lading to the full extent permitted by applicable law, any liability for loss or damage to the goods shall be governed:

"(1) When it is established in whose custody the goods were when the loss or damage to the goods occurred:

(a) During sea-carriage or during carriage by inland waterways (which shall be deemed to be sea-carriage) by the Carriage of Goods by Sea Act of the United States as provided in Clause 2 above;

(b) During rail transportation within the United States as permitted by the Interstate Commerce Commission and according to the joint tariffs on file with the Federal Maritime Commission and the Interstate Commerce Commission, the sea carrier guarantees the performance of the other carriers under the joint tariffs. As to services incident to through transportation to be performed by carriers not subject to the joint tariffs or to be performed outside the United States of America, carrier undertakes to procure as shippers' agent such services as necessary. All such services shall be subject to the usual contracts of persons providing the services. Sea carrier guarantees the fulfillment of the obligations of such persons under the pertinent contracts.

"(2) When it cannot be established in whose custody the goods were when the loss or damage occurred, it shall be conclusively presumed to have occurred during sea carriage and any liability therefor shall be governed as provided in (1)(a) above."

The full implication of this provision can best be obtained by referring to the Carriage of Goods by Sea Act and reading the long list of circumstances and conditions that relieve the water carrier of liability for loss and damage.

The Steamship Conference System

Steamship conferences had their origin in the late 1800's. Today the great majority of the water carriers operating in foreign trade are members of one conference or another. There are also some independent carriers who do not belong to any conferences, but they are in the minority.

As a general rule the conference operates in a single trade, as for example the North Atlantic United Kingdom Freight Conference, and usually in one direction only. However, a water carrier operating to and from many countries may belong to several conferences. The primary objective of the conference is the control of competition but, like the railroad or motor carrier rate bureaus, the conferences perform many services for their members. They see to the establishment of uniform rates, rules and regulations, apportion traffic among member lines, do whatever they can to insure the integrity of the conference members and combat, to the extent possible, the competition of non- member lines. The conference agreements as well as their tariffs must be approved by the Federal Maritime Commission. Like Section 5a (now 10706) of the Interstate Commerce Act, Part 529.1 of Title 46, Code of Federal Regulations, gives the individual members of the conference the right of independent action.

The most controversial feature of the steamship conference has been the dual rate system. Under this system the shipper signs a contract with the conference, usually for a period of one year, agreeing to give conference lines all of his shipments in their particular trade in return for a lower "contract rate."

Part 538.1 of Title 46, Code of Federal Regulations, provides that:

> "Any carrier or conference of carriers desiring to institute a new dual-rate contract system should file with the Commission its proposed form of contract *** requesting approval of the dual-rate contract system and of the form of contract to be used *** ."

The terms of a "Uniform Merchant's Contract" are specified in Part 538.10 of Title 46, C.F.R.

Port Charges and Facilities

Charges assessed at marine terminals and ports of export and import form an integral part of the through transportation charges. These charges may differ considerably depending upon the facilities and services available. The tariffs publishing the regulations and charges should be consulted in determining which port to use. In checking such tariffs it is important to note the weight on which the handling charges are assessed. The transportation charges may be applied on the basis of a minimum weight, but as a general rule the wharfage and handling charges are computed on the actual weight of the shipment. The following paragraphs describe briefly some of the important United States ports and their facilities.

North Atlantic Ports

Baltimore—The port of Baltimore consists of the broad tidal estuary forming the mouth of the Patapsco River, which lies on the west side of Chesapeake Bay, and includes more than 40 miles of deep water frontage along the river and its tributaries. The main ship channel is 39 feet deep. Baltimore is the only North American port possessing two different exits to the sea, being located 150 nautical miles from the Chesapeake Capes and approximately 125 nautical miles from the Delaware Capes via the Chesapeake & Delaware Ship Canal. The port differs in one respect from most other American seaports in that practially all of the large marine terminals used by ocean shipping are either owned or operated by the trunkline and terminal railroads—Chessie System, Canton Railway, Con Rail and Western Maryland Railroad—operating at the port.

Boston—The historic port of Boston, located on Massachusetts Bay, has a water frontage of more than 140 miles, of which seven miles has a depth of 35 feet or more. It has over 250 piers or wharves, providing some 30 miles of berthing space. Three main channels lead into the port from the open sea, with depths ranging from 30 to 40 feet. Con Rail and the Boston & Maine Railroad serve the port, as well as over 500 motor carriers.

New York—The great port of New York has 750 miles of waterfront in the New York Harbor area—460 miles straight waterfront in Greater New York and 290 miles in New Jersey. Entrance to the harbor is through the Ambrose Channel with a depth of 45 feet. There are over 250 berths for ships in the general cargo trade, more than 50 berths for tankers and for ships carrying bulk cargoes. Grain transfer facilities are also available. Some 1,500 harbor craft are used to transfer cargo from railroad terminals and private piers to the steamship piers. Four railroad lines—Con Rail, Chessie System, Delaware & Hudson Railroad and Long Island Railroad—play a heavy role in the movement of freight to, from and through the port and numerous common carrier motor truck lines also serve the commercial cargo needs of the gateway. The port includes Foreign Trade Zone No. 1, which was opened on Staten Island on February 1, 1937.

The Port Authority of New York and New Jersey owns six major marine terminals, three on the Brooklyn waterfront and three on the New Jersey side of the port. The newest and largest of these is the Newark-Elizabeth Marine Terminal, which provides eight miles of ship berthing space and was designed to provide adequate facilities for the trans-shipment of containers.

Philadelphia—Located on the Delaware River, the port of Philadelphia has a channel dredged to a depth of 40 feet from deep water in Delaware Bay. Nearby ports which may be included in the port complex are Marcus Hook and Chester, Pa., Trenton, Camden and Paulsboro, N.J., and Wilmington, Del. This area is served by the Chessie System, Con Rail and Philadelphia Belt Line RR, as well as by hundreds of motor carriers. More than a hundred steamship lines provide service between the port and points all over the world. Lighterage is unnecessary, as cargoes can be loaded and unloaded directly at shipside.

Ports of Hampton Roads include Norfolk, Newport News and Portsmouth, situated in the extreme southwestern corner of Chesapeake Bay. The Chessie System, Norfolk & Western Railroad, Seaboard Coast Line, Southern Railway, Norfolk, Franklin and Danville, and Norfolk and Portsmouth Belt Line serve these ports.

Other ports included within the North Atlantic group are:

Albany, N.Y.
Alexandria, Va.
Fall River, Mass.
Hopewell, Va.
New Haven, Conn.
New London, Conn.
Portland, Me.
Providence, R.I.
Richmond, Va.

South Atlantic Ports

Charleston—A large, natural harbor, the port of Charleston is about 7½ miles from the open sea, with an entrance channel 35 feet in depth.

Port facilities are owned and operated by the South Carolina State Ports Authority. Inland service is provided by the Southern and Seaboard Coast Line railways and numerous truck lines. Steamships operate on regular schedules to principal ports throughout the world.

Savannah—Located on the south bank of the Savannah River, the port of Savannah is the largest of the South Atlantic ports. It is situated about 15 miles from the ocean and has a depth of 34 feet in the main harbor. The ample dock facilities are owned and operated by the Georgia Port Authority. Rail service to and from the port is provided by the Seaboard Coast Line and the Southern Railway System. Adequate truck service is also available and some 30 steamship companies have representatives at Savannah.

Additional South Atlantic and Florida ports are:

Brunswick, Ga.
Georgetown, S.C.
Jacksonville, Fla.
Miami, Fla.
Morehead City, N.C.
Port Everglades, Fla.
Port Royal, S.C.

Gulf Ports

Houston—This port is located on Buffalo Bayou in the southeastern part of Texas at the head of the Houston Ship Channel, a waterway 50 miles long, 300 to 400 feet wide and over 36 feet deep, connecting with the Gulf of Mexico through the Galveston entrance jetties. The port is served by the Southern Pacific, Missouri Pacific, Santa Fe, Missouri-Kansas-Texas, Burlington Lines and Rock Island Lines railroads. The Public Belt Railway connects the trunklines with the port. Some 30 motor carriers are also available for service to and from the port.

Mobile—The State of Alabama's port is located on the west bank of the Mobile River at the head of Mobile Bay, 30 miles from the Gulf of Mexico. Minimum depth of the channels leading to the port is 36 feet. Public terminals facilities are operated by the Alabama State Docks. Railroad service is provided by the Louisville & Nashville, Illinois Central Gulf and Southern Railway railways. In addition to the three trunk lines, the Alabama State Docks operate a terminal railway.

New Orleans—The "Crescent City" port is located on the Mississippi River, 110 miles from the Gulf of Mexico. Facilities include modern quay-type wharves and steel transit-sheds, a public grain elevator and one of the largest public commodity warehouses in the world. Foreign

Trade Zone No. 2, opened in 1947, offers numerous processing facilities. New Orleans is served by seven trunk line railroads—Illinois Central Gulf, Southern Pacific, Missouri Pacific, Southern Railway, Texas and Pacific, Louisville and Nashville, and Kansas City Southern—all of which connect with the port through the municipally owned New Orleans Public Belt Railroad.

The Gulf Ports include:

Baton Rouge, La.
Beaumont, Tex.
Brownsville, Tex.
Burnside, La.
Corpus Christi, Tex.
Freeport, Tex.
Galveston, Tex.
Gulfport, Miss.
Lake Charles, La.
Panama City, Fla.
Pensacola, Fla.
Pascagoula, Miss.
Port Allen, La.
Tampa, Fla.

Pacific Coast Ports

Los Angeles—The port of Los Angeles has 28 miles of water frontage and over 66,000 feet of municipal wharves—all available for the convenience of ships plying the seven seas. The port is divided into three districts—San Pedro, the outer harbor; Terminal Island; and the inner harbor, referred to as the Wilmington District. The harbor is on San Pedro Bay, about 25 miles south of the City of Los Angeles. The water depth at the harbor entrance is 47 feet at low tide, and 35 feet at all other parts of the harbor. The Santa Fe, Southern Pacific and Union Pacific railways all serve the port through the Harbor Belt Railroad, which is the operating agency of the trunk lines in conducting joint freight terminal operation at Los Angeles Harbor.

Portland—Up the Columbia River, 110 miles from the Pacific Ocean, is the port of Portland, tidewater gateway to the producing areas of Oregon, Eastern Washington and Idaho. This "commercial capital of the Great Columbia Basin" is one of the world's great fresh water ports, served by three major railroads—Burlington Northern, Southern Pacific and Union Pacific. The port is joined to the ocean, 110 miles away, via the Willamette and Columbia Rivers by a 35-foot channel. It has harbor frontage of 30 miles and not only handles ocean ships but vessels in the inland waterways trade. Several motor carriers offer convenient service to and from the port.

San Francisco—The port of San Francisco has been serving world commerce for over a century and offers an unsurpassed combination of experience and facilities to shippers and the shipping industry. Its harbor covers over 450 square miles and has a shore line, exclusive of navigable inlets, of 100 miles with a depth of water from 8 to 10 fathoms, an anchorage capable of accommodating any class of ships. The port has 18 miles of berthing space and 195 acres of pier and wharf area. At many piers there is an alongside water depth of 45 feet, sufficient to accommodate the world's largest vessels. The entire waterfront is served by the State Belt Railroad, which acts as a connection between the port and the trunk line railroads—the Southern Pacific, Santa Fe, and Western Pacific railroads. Foreign Trade Zone No. 3 has been in operation in the San Francisco port area since 1948.

Seattle—This North Pacific Coast port is known as the "Port of Opportunity" and the "Gateway to Alaska and the Orient." It is a natural port and its main harbor, Elliott Bay, is an arm of Puget Sound with sufficient depth for the world's largest vessels. The Chicago, Milwaukee, St. Paul & Pacific Railroad, Burlington Northern, and Union Pacific System are the three railroads that serve the port of Seattle. Excellent motor carrier service is also available. A Foreign Trade Zone has been in operation at Seattle since 1949.

The Pacific Coast Ports include:

Anacortes, Wash.
Anchorage, Alaska
Astoria, Ore.
Bellingham, Wash.
Everett, Wash.
Kalama, Wash.
Long Beach, Cal.
Longview, Wash.
Oakland, Cal.
Olympia, Wash.
Port Angeles, Wash.
Port Hueneme, Cal.
Port San Luis, Cal.
Sacramento, Cal.
San Diego, Cal.
Stockton, Cal.
Tacoma, Wash.

Great Lakes Ports

In 1784, George Washington wrote to a member of the Continental Congress: "Open the road (between the Great Lakes and the sea) ... and see what an influx of articles will pour in—how amazingly our exports will be increased—and how amply we shall be compensated for any trouble and expense we may encounter to effect it."

This vision became an actuality with the official opening of the St.

Lawrence Seaway on June 26, 1959. The Seaway is really the fourth seacoast of the nation, linking mid-America with the Atlantic Ocean and the rest of the world. It was built by America and Canada, with 22,000 men working four years, and a cost of $480,000,000. Hydroelectric projects cost another $600,000,000. A tremendous engineering achievement, but even more—an example of the closest cooperation between two great and friendly nations.

The St. Lawrence Seaway is capable of handling 90 percent of the world's merchant vessels. The principal ports served by the Seaway include:

- Ashtabula, Ohio
- Buffalo, N.Y.
- Chicago, Ill.
- Cleveland, Ohio
- Detroit, Mich.
- Duluth-Superior, Minn.-Wis.
- Erie, Penn.
- Green Bay, Wis.
- Hamilton, Ont.
- Milwaukee, Wis.
- Muskegon, Mich.
- Oswego, N.Y.
- Port Huron, Mich.
- Toledo, Ohio
- Toronto, Ont.

A great variety of railroads and motor carriers serve these cities and the shipping lines that call at the ports offer service to most of the world. The port of Toledo area includes a Foreign Trade Zone.

CHAPTER 2

IMPORT AND EXPORT TARIFFS AND RATES

General

The subject of import and export tariffs and rates will be presented in this chapter in the light of their relation to inland traffic only. In other words, just as the jurisdiction of the Interstate Commerce Commission over foreign commerce is limited to the movement between the point of origin or destination in the United States and the port of trans-shipment, and it must treat the import and export rate situation as though the ports were destinations instead of gateways, the present discussion will be devoted primarily to those features of import and export rates which apply to the domestic carriers operating to or from the port, and not to the ocean carrier which operates beyond the ports. Of necessity, however, some consideration must be given to the subject of through, intermodal import and export rates.

Jurisdiction of Interstate Commerce Commission

As soon as foreign commerce is brought through a port of entry in the United States and is accepted by an I.C.C. regulated carrier for movement to a destination in the United States it becomes subject to the provisons of the Revised Interstate Commerce Act (Public Law 95-473) and the jurisdiction of the Interstate Commerce Commission. Likewise, traffic moving from an inland point to a port of exit for shipment to a foreign country is similarly regulated.

At this stage of our studies there can be no doubt as to the jurisdiction of the Commission with regard to interstate movements by railroad, motor carrier, domestic water carrier or freight forwarder. However, where foreign commerce is concerned we encounter a slightly different situation. Where shipments originate in the United States and are destined to points in Canada the Commission's jurisdiction extends only to that portion of the movement which takes place in the United States. The carriers may join with the Canadian lines in through rates but the I.C.C. has no power to prescribe such joint international rates and, should the Commission find such joint through rates unreasonable, they may issue an order against the United States carrier only.

The jurisdiction of the Commisson does not extend to movements from points in a foreign country through the United States to points in a foreign country. This freedom from regulation of traffic moving through the United States has had little significance in the past, however, the development of the "land-bridge" concept gives greater importance to movements of this kind.

In *Consolidated Truck Lines v. Fess & Wittmeyer,* 83 MCC 673, the Commission said:

> "Congress, by the 1954 Amendment of Section 203(a)(11), did not intend to restrict, but rather to broaden, the scope of the Commission's jurisdiction in certain respects to embrace foreign carriers engaged in transportation in this country of shipments which have neither origin nor destination in the United States."

In passing upon and interpreting its jurisdiction over import and export traffic, the Commission, in *E. Fernandez & Co. v. S.P. R.R. Co.,* 104 I.C.C. 193, 194, stated:

> "Section 1 of the Act provides:
>
> "That the provisions of this Act shall apply to common carriers engaged in the transportation of passengers or property *** from one State or Territory of the United States, or the District of Columbia, to any other State or Territory of the United States, or the District of Columbia, or from one place in a Territory to another place in the same Territory, or from any place in the United States through a foreign country to any other place in the United States, or from or to any place in the United States to or from a foreign country, but only insofar as such transportation *** takes place within the United States.
>
> "It will be noted that transportation from a foreign country through the United States to a foreign country is not embraced in this section.
>
> "Since March 2, 1917, *** the act has not been applicable to Puerto Rico. Neither does the act confer upon us jurisdiction over the transportation of property from Mexico through the United States to Puerto Rico, insofar as such transportation takes place within the United States, for even though the transportation from Mexico through the United States to Puerto Rico were shown to have been partly by railroad and partly by water under a common contol, management, or arrangement for a continuous carriage or shipment *** and even though Puerto Rico is a territory appurtenant to the United States we would have no jurisdiction over such transportation. To hold otherwise would be to extend the act contrary to the provisions above cited.' "

This principle had previously been discussed by the Commission in *John S. Seymour v. M.L. & T. R.R. & S.S. Co.,* 35 I.C.C. 492, 493, as follows:

> "The sugar was transported from a nonadjacent foreign country, through the United States, to destinations in an adjacent foreign country. We entertain no doubt that the regulatory power of Congress extends to the transportation within this country, but apparently the jurisdiction of this Commission does not."

In the case of the *Chamber of Commerce of the State of New York, et al, v. N.Y.C. & H.R. Railroad Company,* 24 I.C.C. 55, 74, in further clarification of the extent of its jurisdiction over foreign commerce, the Commission stated:

> "We have no jurisdiction of the ocean rates and must deal with the import and export rate situation as though the ports were destinations instead of gateways."

As the Interstate Commerce Act, therefore, grants to the Commission jurisdiction over only that portion of the export or import movement which takes place within the United States, it is readily apparent that the other provisions of the Act as to justness and reasonableness, undue discrimination, undue prejudice and filing of tariffs are enforceable by the Commission on only such portion of the movement as takes place within the United States.

In *Central Vermont Ry. v. United States,* 182 Fed. Supp. 516, the court ruled:

> "Rail carriers within the United States have a clear alternative with respect to their participation in international traffic. They may publish rates to or from the international boundary, which apply only to that part of the transportation which they themselves perform; or they may join with the foreign carrier in publishing a joint rate for the entire transportation, thus becoming jointly and severally liable for any unlawfulness in such rates."

The Supreme Court of the United States, in *Porter Co., Inc. v. Central Vermont Ry.,* 366 U.S. 272, said:

> "The Commission's jurisdiction over transportation from Canada to points in the United States extends only to that part of such transportation which takes place within the United States. Thus in issuing an order directing defendant railroads, 'according as they participate in the transportation within the United States' to cease their participation in practices found to be in violation of Section 3(1), the Commission did not attempt to control the Canadian portion of considered transportation; nor did it exceed its jurisdiction."

With regard to joint, through, intermodal export and import rates the jurisdiction of the Interstate Commerce Commission is limited to that portion of the movement within the United States. The fact that the ocean carrier may have entered into an arrangement with the inland carrier in the United States under which the through rate is published in the domestic carrier's tariff would not subject the ocean carrier to the jurisdiction of the I.C.C. Even though the through tariff is filed with the I.C.C. by the ocean carrier or conference of ocean carriers as now provided for in the Commission's regulations, the ocean carrier is not subject to the jurisdiction of the Interstate Commerce Commission.

The Through Export Bill of Lading

In 1920 Section 25 was added to the Interstate Commerce Act requiring the Interstate Commerce Commission to prescribe a form of through bill of lading for use by railroads and steamship lines on export shipments. Such a bill of lading was prescribed by the Commission in Export Bill of Lading, 64 I.C.C. 347 (1921). Section 25 of the Act was repealed in 1940, but it was not until 1966 that the Federal Regulation prescribing such a form was revoked.

Considerable study has been given to the need for some form of shipping document that would satisfy the needs of shippers and carriers alike. If and when a standard international, intermodal, joint, through bill of lading is developed and approved by the regulatory agencies it will, in all probability, be a composite of the ocean and inland bills of lading now in use. The face of the bill of lading should provide all of the information concerning the shipment, show freight rate and charges, specify routing and on the reverse side the contract terms and conditions. Such terms must, of necessity, include the requirements of Section 11707 of P.L. 95-473 as well as the pertinent provisions of the Carriage of Goods by Sea Act.

It should, of course, be understood that although more and more through bills of lading will be used in the future, there will continue to be many international shipments moving on separate bills of lading, one covering the movement to the port and one or more beyond.

Publication of Import and Export Rates

On traffic between the United States and an adjacent foreign country (Mexico or Canada), the Commission has not required the establishment of separate rates to the border point to cover that portion of the haul within the United States, but has considered publication and filing with it of the joint through rates between the American and Canadian or Mexican carriers sufficient to meet the requirements of the Act.

In commenting upon the situation in the light of the decision of the Supreme Court of the United States in the case of *News Syndicate Company v. N.Y.C.R.R. Co., et al,* 275 U.S. 179, the Commission, in its forty-second annual report to Congress, stated as follows:

> "Section 1 of the Interstate Commerce Act makes the provisions of the act applicable to common carriers engaged in the transportation of passengers and property wholly by railroad 'from or to any place in the United States to or from a foreign country, but only insofar as such transportation *** takes place within the United States,' and confers upon us jurisdiction over such transportation and over the carriers therein engaged. Also, paragraph (1) of Section 6 contains, among other things, provisions which are as follows:
>
> > 'That every common carrier subject to the provisions of this Act shall file with the Commission created by this Act and print and keep open to public inspection schedules showing all the rates, fares, and charges for transportation between different points on its own route and between points on its own route and points on the route of any other carrier by railroad, *** when a through route and joint rate have been established. If no joint rate over the through route has been established, the several carriers in such through route shall file, print and keep open to public inspection as aforesaid, the separately established rates, fares, and charges applied to the through transportation. The schedule printed as aforesaid by any such common carrier shall plainly state the places between which property and passengers will be carried. *** The provisions of this section shall apply to all traffic, transportation and facilities defined in this Act.'
>
> "In the past these provisions of Section 6 have been interpreted by carriers operating within the United States and by us as leaving the carriers free, in connection with transportation from and to the United States to and from Canada and Mexico, either to publish and file separately, in each instance, a rate for the transportation which takes place within the United States, or, in conjunction with carrier operating either in Canada or in Mexico, a joint rate covering transportation over the entire route. Acting in accordance with this interpretation, carriers operating within the United States, generally speaking, have heretofore published and filed in connection with the transportation mentioned, only joint through rates.
>
> "On November 21, 1927, however, in its decision in *News Syndicate Company v. New York Central Railroad Company, et al,* 275 U.S. 179, the Supreme Court held it to be the duty of carriers subject to the act to publish and file, in each instance, a rate applicable only to that part of the

transportation which takes place within the United States. This case involved the question of the validity of a reparation order made by us in *Ontario Paper Company, Limited, et al, v. Canadian National Railways, et al,* 95 I.C.C. 66, 102 I.C.C. 365, wherein we found to be unreasonable joint rates applied by defendant carriers to the transportation of newsprint paper from Thorold, Ontario, to New York City, and awarded reparation against carriers operating within the United States for the difference between the charges collected by them under the joint rates and the charges we found would have been reasonable for the transportation services covered by those rates. Both in the proceeding before us and in the case in court it was shown that no rate had been published or filed which was applicable only to the transportation which took place within the United States. On behalf of the carriers it was contended that, under the circumstances described, we are without jurisdiction either to pass upon the reasonableness of the joint rate, or to make an order of reparation if we find the joint rate to be unreasonable, and in holding these contentions to be without merit the Supreme Court, among other things, said:

> ' *** The Interstate Commerce Act applies to the lines that carried, and to the transportation of, the paper from the international boundary to New York City. *** It was the duty of defendants in error to establish just and reasonable rates for that service. *** They failed to make or publish any rate applicable to that part of the transportation. Section 8 makes them liable for damages sustained in consequence of such failure. Had the through rate been just and reasonable, no damages would have resulted to plaintiff in error. Its right to reparation does not depend upon the amounts retained by defendants in error pursuant to agreed divisions. *** Their breach of the statutory duty was a proximate cause of the losses complained of. The failure to establish rates covering the transportation from the international boundary contravened the provisions of the Act and compelled plaintiff in error to pay the through charges complained of. The Commission had jurisdiction to determine whether plaintiff in error was entitled to an 'award of damages under the provisions of this Act for a violation thereof.' *** And it was the duty of the Commission to ascertain the damages sustained. It is obvious that, in the ascertainment of damages, the Commission had jurisdiction to determine the reasonableness of the charges exacted.'

"Following the rendition of the decision mentioned we received a large number of inquiries concerning the effect of the decision upon commerce, particularly commerce between points in the United States and points in Canada, whereupon we instituted a proceeding, Ex parte 93, and invited a full expression of views in the premises by interested parties. After briefs had been filed the proceeding was assigned for the hearing of oral argument, which was held on October 11, 1928. At that argument, and in briefs filed in our office, it was developed that there are many and material differences between the views entertained by counsel for the carriers on the one hand and by those who represent the shipping and traveling public on the other, but they were all agreed that nothing should be done which would result in bringing about a cancellation of the joint through rates. It was shown that the volume of traffic covered by such rates is very large, and that the number of joint rates involved is correspondingly great. Attention was called to the fact, in many instances, points of origin and points of destination of the traffic are grouped, and that average mileages instead of actual mileages are used in establishing rates to and from points in the groups and in agreeing upon the divisions of the joint rates as between carriers operating in the United States on the one hand and those operating in Canada and Mexico on the other. For these reasons, among others, it was stated that, as a practical matter, it is impossible for the carrier to publish and file in each instance a rate applicable only to that part of the transportation which takes place within the United States. It was also stated that, even assuming that rates confined to the transportation which takes place within the United States can be established, their substitution for the joint through rates now in force will result in great inconvenience and cause much unnecessary expense to all interested parties, including the carriers who perform the transportation services and the members of the general public for whom the services are performed.

"Like views concerning the undesirability of canceling the joint through rates were included in communications received by us subsequent to the date of the hearing of oral argument above mentioned, and with one of the communications was enclosed copy of an extract from a Canadian paper, which by way of illustration, we set forth as follows:

> 'Trade between Canada and the United States will be seriously disturbed if the Interstate Commerce Commission, which meets in Washington on October 11, prohibits through freight rates between the two countries. The question has arisen over a suit brought by the News Publishing Company, of New York, against the New York Central Railroad, which was decided by the Supreme Court of the United States in November, 1927. It concerned shipments of paper from Canada to the United States under joint through rates, and the court held that the United States carriers had failed to make public the rates applicable to their part of the total haul, and consequently were liable in damages. Apparently the opinion of the Court was that international freight must be carried not on through rates, but, so far as the United States is concerned, on the domestic rate from the border. In view of the importance of the case, the United States Interstate Commerce Commission, at its meeting next week, will hear argument on the question whether it can properly sanction international rates, whether such rates should be published in the United States, and whether there should be similar publication for the part of the haul in Canada. Mr. McKeown, chairman of the

board of Railway Commissioners of Canada, was invited to sit with the American Commission in Washington, but regretted that the Board is not in a position to be represented at the present time. If the through rates are prohibited, the effect will be felt by a trade which amounted in the aggregate to $1,128,787,658 during the past calendar year. But the exports of the United States to Canada were $215,684,622 greater than the exports of Canada to the United States, which will suffer most from any such change, and it is likely that the Interstate Commerce Commission will find some way of avoiding it.'

"But while, as above stated, all interested parties appear to be anxious that the joint through rates shall be continued in force, there is a material difference between the views expressed by counsel for the carriers and those imparted by representatives of shippers concerning the jurisdiction which should be exercised by us in connection with such rates. Counsel for the carriers urge that in passing upon the question of unreasonableness prohibited by Section 1 of the Act we should make findings which will affect only rates to be applied in the future, and should decline to find unreasonable, and to award reparation in connection with, any joint through rate which may have been applied to and collected for, transportation prior to the effective dates of any findings or orders we may make. On the other hand, representatives of shippers and consignees insist that we should continue to award reparation against carriers operating in the United States, in each case where we find to have been unreasonable the joint through rate so applied and collected."

As far as through joint rates in connection with ocean carriers to points in foreign countries are concerned, the situation is considerably different. In the first place, the filing of joint through rates for movements via domestic rail, motor or water carrier and ocean carriers in international trade involves compliance with the regulations of two federal agencies, the Interstate Commerce Commission and the Federal Maritime Commission. The tariff filing regulations authorized by the Interstate Commerce Act for railroads, motor carriers, domestic water carriers and freight forwarders are contained in Parts 1300, 1307, 1308 of the Code of Federal Regulations and those applicable to common carriers by water and conferences of such carriers in the foreign commerce of the United States are found in Part 536 of the Code of Federal Regulations.

Although there is considerable similarity in the tariff filing regulations of each agency, there are sufficient differences to create problems. On July 31, 1969 the Interstate Commerce Commission instituted a proposed rulemaking proceeding (Ex Parte 261) with a view to amending its existing tariff rules pertaining to export and import traffic moving via rail, motor or water common carrier in conjunction with oceangoing water common carriers operating in foreign commerce.

In their first report, issued on September 4, 1970, their findings were:

> "We find that we have jurisdiction to prescribe and to establish reasonable rules and regulations governing the filing, posting and publishing of tariffs containing both joint and combination rates for movement of international traffic over through routes."

They also stated in their report:

> "We wish to make clear, however, that we are not deciding herein that the ocean carrier participants in international joint rates come under our jurisdiction. Those carriers remain subject to the jurisdiction of the Federal Maritime Commission even when joint rates are established. Though this agency is authorized to accept for filing and to regulate the joint rates to the extent transportation occurs within this country, we are not thereby empowered to preempt any of the statutory duties that have been conferred by Congress on the Federal Maritime Commission. Accordingly, nothing stated herein should be construed as having the effect of barring that Commission from directing the participating water carriers to file tariffs with it, at the same time joint rate tariffs are filed with us, so that it can properly discharge its duties."

In their final report in Ex Parte 261, the Commission said:

> "Every tariff naming such a through route and joint rate shall be filed with this Commission. The tariff may be filed in the name of the ocean carrier, a conference of ocean carriers, the domestic carrier or the duly appointed tariff publishing agent of such carriers."

This proceeding (Ex Parte 261) and the specified rules prescribed by the Commission will be discussed in more detail in a subsequent chapter.

Export and Import Rates Versus Domestic Rates

Export and import rates when so designated, whether class or commodity, take precedence over other rates between the same points, via the same route on export or import traffic as the case may be, except when and insofar as alternative use of other class and commodity rates is specifically provided in the tariff containing such export and import rates. Furthermore, an export rate established by an intermediate rule takes precedence over a domestic rate (*Nueces County Nav. District v. A.T. & S.F.*, 315 I.C.C. 155).

In *Louisiana & A. Ry. v. Export Drum Co.*, 228 F. Supp. 89, the court held:

> "All of the requisites of the tariff must be met in order to entitle shipper to an export rate. Defendant's failure to comply with all the tariff conditions results in application of the higher domestic rates on its shipments."

In the absence of specific export or import rates, domestic rates are applicable on export or import traffic. While rates published to apply only on export or import traffic are generally found to be lower than domestic rates between the same points, the carriers are not required to publish reduced rates. The Commission brought this out very clearly in *Central Pennsylvania Coal Producers Assn. v. B.&O. R.R. Co.*, 196 I.C.C. 203, 236, in the following manner:

> "There is nothing inherent in export traffic as such that entitles it to rates lower than on domestic traffic, but the rail carriers are at liberty, if they so desire, to establish rates on export traffic lower than on domestic traffic provided no violation of the act results."

A domestic rate cannot be attacked as being unreasonable just because it is higher than the export or import rate. The reason for this was given by the Commission in *Sumter Packing Co. v. A.C.L. R.R. Co.*, 157 I.C.C. 137, 142, wherein they stated:

> "There are circumstances and conditions generally surrounding the making of import and coastwise rates which do not affect the domestic rates."

This principle was again touched upon by the Commission in *Dallas Cotton Exchange v. A.T. & S.F. Ry. Co.*, 163 I.C.C. 57, 60, as follows:

> "Complainant's contention that the rate assailed is unreasonable when compared with the import rate has little weight, for, as we have heretofore stated, import rates are frequently lower than domestic rates, the former being influenced by considerations which are unrelated to, and have little, if any, bearing upon the reasonableness per se of the domestic rates."

Following this line of thought, we arrive at the logical conclusion that an export or import rate may be higher or lower than the domestic rate depending upon the circumstances and conditions.

Port Differentials

In studying the rate situation in connection with import and export traffic, an understanding of the port differential adjustment is essential. This adjustment is of long standing and has been the subject of much controversy. It is not our purpose to set forth the early history of this subject, but rather to record the most essential facts of the port relationship so that one will have sufficient understanding of this subject to deal with a practical situation.

North Atlantic Ports

Prior to the Interstate Commerce Commission's decision in *Eastern Class Rate Investigation,* 164 I.C.C. 314, the differentials applicable to North Atlantic ports had remained unchanged for a number of years. Rates to and from New York were the basic rates. Rates from central territory to Philadelphia, Pa., and Baltimore, Md., were uniformly, on all classes, 2 and 3 cents respectively, lower than New York. Westbound the same differentials applied except that on first and second classes they were 6 cents at Philadelphia and 8 cents at Baltimore. The domestic rates generally were the same as the export rates. Other North Atlantic ports were related to the same adjustment—Boston, Mass., same as New York rates; Norfolk, Va., same as Baltimore rates; Canadian ports, same as Baltimore rates on import traffic; and on export traffic, Montreal same as Philadelphia rates and other Canadian ports same as New York rates.

Effective December 3, 1931, the Commission, in its decision in the *Eastern Class Rate Case,* prescribed domestic rates within and between trunk-line and central territories on a distance basis, which had the effect of destroying the long-established port differential relations on domestic traffic, the decision not applying to import and export rates. As a result of this adjustment the new domestic port rates generally were less than the existing port rates, so carriers found it necessary to reconstruct the latter. In establishing the new port rates the domestic rates to and from Baltimore were used as the basis and the former differentials were added thereto to make rates to and from Philadelphia and New York, except where, as in the northern portion of central territory, that basis would produce rates to and from the latter ports higher than the corresponding domestic rates. In such instances the domestic rates to and from New York generally were used as the basic rates and the differentials were deducted therefrom to arrive at port rates to and from Philadelphia and Baltimore. For example, the domestic sixth class rates from Detroit, Mich., were 33 cents to Baltimore and 34 cents to New York. Adding the 3-cent differential to the Baltimore rate would produce a port rate to New York of 36 cents, or 2 cents higher than the domestic rate. In that instance the port differential was deducted from the New York domestic rate to arrive at a Baltimore port rate of 31 cents, or 2 cents less than Baltimore domestic rate. The rates to and from Norfolk, Boston, and Canadian ports were continued on the same relation as formerly to Baltimore, Philadelphia, or New York.

As a result of this adjustment, which became effective January 3,

1932, the port rates as a whole became lower and in a number of cases substantially lower than the corresponding domestic rates. Port commodity rates lower than the corresponding domestic rates were established on certain articles at the ports and were related differentially in the same manner as the class rates.

Port of Albany

Early in the history of the country Albany was an important port, but as steamships and steel hull vessels, whose greater draft would not permit navigation of the Hudson River, gradually superseded the sloops and schooners, Albany's ocean commerce was almost wholly diverted to other ports. In 1925, the New York State Legislature created the Albany Port District and provided for the appointment of the Albany Port District Commission. Improvements in the port terminal facilities were made, and with the widening and deepening of the river, authorized by Congress in 1915 and completed in 1934, the Port of Albany was again a deepwater port which could be reached by 85 percent of the ocean-going ships in regular freight service and in competition with other North Atlantic ports.

Albany was not related to the other ports by fixed differentials, the full domestic rates applying on Albany's port traffic. Upon investigation, by complaint of the Albany Port District Commission, the Interstate Commerce Commission in *Albany Port District Comm. v. Ahnapee W. Ry. Co.,* 219 I.C.C. 151, concluded that maximum reasonable port rates between the port of Albany and points in central and western territories would be rates the same in amount as the port rates concurrently applicable between the port of Philadelphia and the same points, subject to Albany's corresponding domestic rates as maxima.

North Atlantic Port Differentials Disapproved by Court

In *I. & S. No. 6615, Equalization of Rates at North Atlantic Ports,* 311 I.C.C. 689, the railroads serving the northern tier ports of Portland, Me., Boston, Mass., Albany, N.Y., and New York, N.Y., had published reduced export-import rates between those ports and Central Territory reflecting the same level of rates as were in effect between Central Territory and the southern tier ports of Baltimore, Md., Philadelphia, Pa., and Hampton Roads. The justification given for the adjustment was that over the years the relative share of tonnage handled through the northern tier ports had declined in comparison with that going through the southern tier ports. The railroads serving the latter ports retaliated by publishing reduced rates to restore the differentials described earlier in this chapter, although the carriers indicated that if the reduced rates to the northern ports should be found unlawful, the new rates to the southern ports would be cancelled voluntarily. Both sets of rates were suspended and investigated by the Commission, which in June, 1961, concluded that the reduced rates to and from the northern tier ports were unduly preferential of those ports and unduly prejudicial to the southern tier ports in violation of Section 3(1) of the Interstate Commerce Act, and ordered the rates cancelled. The retaliatory adjustment to and from the southern tier was found not to have been shown just and reasonable and ordered to be cancelled, also. The net result of the investigation proceeding was to maintain the existing differential relationship.

In February, 1962, however, the Commission's order was set aside by a three judge court on the ground that it was erroneous in law and lacked "the rational basis necessary to uphold it." The court held that the Commission had interpreted Section 3(1) of the Act too narrowly and failed to give sufficient weight to the interests of shippers, receivers, carriers and inland localities. On appeal to the U.S. Supreme Court, the Commission contended that the lower court's ruling would permit destructive competition inconsistent with the encouragement of sound economic conditions among carriers, as required by the National Transportation Policy. Railroads serving the southern tier supported the Commission, claiming that elimination of the differentials would give New York a monopoly on export-import traffic. The Justice Department and the railroads serving the northern ports opposed the appeal.

On May 20, 1963, the Supreme Court announced a per curiam decision in the appeals cases—*No. 97, Baltimore & Ohio Railroad Co., et al, v. Boston & Maine Railroad, et al; No. 98, Maryland Port Authority v. Boston & Maine Railroad, et al; and No. 99, United States of America v. Boston & Maine Railroad, et al.* With Justice White not participating, the other members of the court were equally divided. The effect of the 4-4 vote was to affirm the judgment of the lower court which had set aside the Commission's order. On the basis of the Court's decision, it is no longer necessary for the railroads to observe a differential relationship among the North Atlantic ports.

There are no port-related rates applying from points east of the Buffalo-Pittsburgh line to the North Atlantic ports.

South Atlantic and Gulf Ports

Generally speaking, prior to the latter part of the year 1919, there were no joint through export rates from central and western territories to the South Atlantic and Gulf ports. Proportional rates applied from certain Ohio River crossings to New Orleans and Mobile, for export, except to Cuba, Europe, Asia, Africa, Australia, New Zealand, and the Philippine Islands. Those rates brought about an equalization of rates from points of origin in central and western territories to New Orleans with the joint through rates from the same points to New York. The first joint through export rates to South Atlantic and Gulf ports were established in December of 1919, by the Director General of Railroads. The rates established were the same as the rates on like traffic to New York and only applied from points in central territory taking 100 percent of the New York-Chicago scale of rates.

For many years the South Atlantic and Gulf port import rates on traffic from Europe, Asia, and Africa have been made differentials under the rates on like traffic from New York to the same destinations. These differentials, known as the Todd-Knott award, applied only to import traffic from Europe, Asia and Africa to points in Central Territory taking 100 percent or more of the New York-Chicago scale. They were governed by the Official Classification and were the following amounts under the standard all-rail domestic rates from New York.

Classes	1	2	3	4	5	6
Cents per 100 pounds	18	18	12	8	6	6

In the case of commodity rates, where the rate on a commodity exceeded the sixth-class rate, the differentials under the rate from New York would be the same as the differential on the class on which the rate was nearest such commodity rate; where the commodity rate was exactly intermediate between two class rates, the differential on the lower class rate would be used; and on commodities taking rates the same as sixth class or lower the sixth-class differential would apply.

Cancellations of import rates from North Atlantic ports in 1918 by the Director General of Railroads, general increases in rates under *Increased Rates, 1920,* 58 I.C.C. 220, i.e., 40 percent in rates from North Atlantic ports and 33⅓ percent in rates from the southern ports and changes from time to time in rates from one port without corresponding changes from other ports, produced differentials more favorable to the Gulf lines than those fixed in the Todd-Knott award. Southern carriers in 1930 filed fourth section applications proposing a readjustment of the rates to restore proper relationship in port rates with North Atlantic ports. This readjustment, with some exceptions, was approved by the Interstate Commerce Commmission in *Export and Import Rates,* 169 I.C.C. 13. In 1934, southern carriers again proposed a new basis of export and import class and commodity rates between the South Atlantic ports of Wilmington, N.C., Charleston, S.C., Savannah and Brunswick, Ga., and Jacksonville, Fla.; the Florida ports of Key West, Miami, Tampa, and Port Tampa; and Gulf ports on the Gulf of Mexico from Pensacola, Fla., to Galveston, Texas, both inclusive, on the one hand and various points in the interior territory, on the other. This new basis of rates, with some exceptions, was approved by the Commission in Docket 3718, *Export and Import Rates To and From Southern Ports,* 205 I.C.C. 511.

While the measure of the rate was different in the foregoing adjustments, the same differential basis was applied and still exists in the present port rates. Generally speaking, this differential basis is as follows:

Import Rates From Gulf Ports—To Territory A—points in central and western territories to the Missouri River—from all foreign countries: Rates are to be the same as those contemporaneously applicable from New York, N.Y., observing the Philadelphia, Pa.-Cincinnati, Ohio rates as minima.

To Territory B—points in central and western territories not included in Territory A, also upper Peninsula of Michigan—on traffic from Europe and Africa: Rates are to be made the Todd-Knott differentials under the rates contemporaneously applicable from New York observing the Norfolk, Va., rates or rates made the following differentials under the rates from Baltimore, Md., whichever are lower, as maxima:

Classes	1	2	3	4	5	6
Cents per 100 pounds	10	10	9	5	3	3

Commodities taking rates lower than sixth-class—3 cents.

To Territory B, on traffic from countries other than those in Europe and Africa: Rates are to be the same as those from Baltimore or Norfolk, whichever are lower.

To Territory C—west of the Missouri River—on traffic from all countries: Rates from all Gulf ports not higher than the lowest domestic rates applicable from any Gulf port, subject to the port relationships established on domestic traffic in *Galveston Commercial Assoc. v. G.H. & S.A. Ry. Co.,* 128 I.C.C. 349; 132 I.C.C. 95. In this case the Commission held that from or to interior points from which the short-line distances to New Orleans do not exceed those to Galveston by more than approximately 25 per cent, rates to or from Galveston and the other Texas ports taking the same rates, as the case may be, shall not exceed the contemporaneous rates on the same commodities to or from New Orleans; and from or to interior points from which the short-line distances to New Orleans exceed those to Galveston and the other Texas ports taking the same rates, as the case may be, shall not exceed rates which are less than the contemporaneous rates on the same commodities to or from New Orleans by differentials of from one to eight cents on the respective commodities named.

Import Rates From South Atlantic Ports—To territories A and B: Rates are to be the same as those from the Gulf ports except that on traffic from countries other than those in Europe and Africa, the rates to St. Louis, Mo., and points taking the same rates will not be higher than the rates from South Atlantic ports to Chicago and the rates to Cairo, Ill., will be the same as from the latter ports to St. Louis.

Export Rates to Gulf Ports—From Territory A to all foreign countries, except Canada, Newfoundland and Nova Scotia, but including insular possessions of the United States and the Panama Canal Zone: Rates are to be the same as those contemporaneously applicable to New York, observing the Cincinnati-Philadelphia rates as minima.

From Territory B to Europe, Africa and east coast of South America:

From Chicago, Indianapolis and Cincinnati: Rates are to be made the differentials stated in the Todd-Knott award, under the rates contemporaneously applicable to New York.

From Louisville, Ky., Evansville, Ind., Cairo and St. Louis, rates are to be made the same as the rates from Cincinnati; and from other points in the territory, rates are to be made the Todd-Knott award differentials under the corresponding rates from said points of origin to New York.

From Territory B to all foreign countries other than those in Europe, Africa and east coast of South America, except Canada, Newfoundland and Nova Scotia, but including insular possessions of the United States and the Panama Canal Zone:

From Chicago and Indianapolis, Ind., rates are to be the same as corresponding rates from said points to Baltimore.

From Cincinnati and Louisville, the same as Norfolk.

From Evansville, Cairo and St. Louis, the same as from Cincinnati.

From all other points the same as rates to Baltimore.

From Territory C to all foreign countries except Canada, Newfoundland and Nova Scotia: The lowest domestic rate to any Gulf port will be published as the maximum export rate to all Gulf ports, Galveston to Pensacola, inclusive, except as provided in *Galveston Commercial Assoc. v. G.H. & S.A. Ry. Co.*, 128 I.C.C. 349; 132 I.C.C. 95 (see above).

Export Rates to South Atlantic Ports—From Territory A and B the same as published to the Gulf ports except:

From St. Louis and Cairo same rates as to Norfolk but not higher than from Chicago to South Atlantic ports.

Export and Import Rates Via Key West, Tampa and Port Tampa—On traffic to and from Territories A, B, and C, rates made by the following differentials over the corresponding rates to and from New Orleans:

Classes	1	2	3	4	5	6
Cents per 100 pounds	62.5	52.5	43	34	29.5	25

Pacific Coast Ports

For many years it has been the practice of transcontinental carriers in meeting competition through the Atlantic ports to accept on export and import traffic moving through the Pacific ports something less than their local or domestic rates. Export and import rates are the same in amount to all of the principal Pacific coast ports, including San Francisco, Calif., and Seattle, Wash. and, while there are exceptions, this is also generally true in respect to the level of domestic rates. Where no special export and import transcontinental rates are provided in T.C.F.B. 3029 series and 3030 series, respectively, the published domestic rates will apply, subject to rules and regulations in export and import rail tariffs.

Transcontinental Eastbound Domestic Tariff 3002 series contains a special export rate section, with lower basis export rates on the principal commodities exported from the Pacific Coast area to markets normally possible to serve through Gulf or Atlantic ports. Similarly, Westbound Domestic Tariff 3001 series contains a section with special import rates on commodities which are received at Gulf or Atlantic ports and move from those ports by rail to the Pacific Coast. These export and import rates apply on traffic to or from foreign countries, insular possessions of the United States, and the Panama Canal Zone.

Purpose of Port Differentials

The purpose of port differentials is twofold: First, to permit carriers operating from and to the different ports to secure some of the export and import traffic which they would not obtain except for the differentials and, second, to distribute the traffic between the different ports. In touching upon this feature, the Commission, in the case of the *Port of New York Authority v. Baltimore & O. R. Co.*, 248 I.C.C. 165, 179, stated:

> "That the several ports have a right to participate in export traffic and that the public interest requires that this right shall be recognized, was a principle accepted by the Commission in *In re Differential Rates, supra,* (11 I.C.C. 13) at page 75, and the legislative history of the amendment of Section 3 of the Act in 1935, bringing ports and port districts within the scope of that section, clearly indicates that fair distribution of export traffic through the various ports, so far as that can be effected through lawful rate adjustments, was one of the contemplated purposes of the amendment. This was recognized by the Commission in *Albany Port District Comm. v. Ahanpee & W. Ry. Co.*, 219 I.C.C. 151, where at page 173, the following excerpt from the report of the House Committee is quoted:
>
>> "In recommending that this bill be passed, the committee does so with the idea in mind that by amending Section 3 of the Interstate Commerce Act as thus contemplated it will encourage and promote the freedom of movement of export, import, and coastwise commerce through the ports of the country. The committee considers that it is to the interest of the public that such commerce be permitted to move freely through as many available ports as the governing circumstances will reasonably permit, and that no restrictions upon and impediments to the free movement thereof should be imposed that are not clearly shown to be sound or economically justified. The recomendation of the committee that this bill be enacted is intended to afford competing ports a forum in which to complain of rate adjustments which tend to concentrate the movement of the traffic through one port or a limited number of ports and to deprive other ports of an opportunity to handle a part of such traffic. The committee believes that such a diffusion of the traffic which moves through the ports will rebound to the benefit of the producer and consumer in the interior by whom in the last analysis the transportation charges levied both for the transportation thereof and for the use of the facilities at the ports are ultimately borne."

Motor Carrier Export-Import Rates

As a rule, motor carriers are selective in the establishment of special export-import rates. Such rates are designed to attract specific movements and it is not necessary to equalize rates among the various ports. Since the number of motor carrier export-import rates is relatively small, they are often included in tariffs naming domestic rates. When this is done, items containing the export-import rates bear a notation limiting their application to export-import traffic. In a few instances, the volume of export-import rates has justified the publication of tariffs devoted exclusively to such rates. Traffic which does not enjoy special export-import rates is subject to domestic rates, and, in addition, may have to bear extra charges for pier pick-up and delivery, reflecting the high cost of providing service at congested piers and wharves.

CHAPTER 3

THROUGH ROUTES

General

In *Thompson vs. U.S.,* 343 U.S. 549 the Supreme Court defined the terms "through routes" and "joint rates" as follows:

> "A 'through route' is an arrangement, express or implied, between connecting railroads for the continuous carriage of goods from the originating point on the line of one carrier to destination on the line of another. Through carriage implies a 'through rate.' This 'through rate' is not necessarily a 'joint rate.' It may be merely an aggregation of separate rates fixed independently by the several carriers forming the 'through route' as where the 'through rate' is the 'sum of the locals' on the several connecting lines or is the sum of lower rates otherwise separately established by them for through transportation."

Two or more carriers may join together in a through route and publish a through rate to cover, and by such procedure set up an entirely new and continuous artery of commerce which is considered as a single unit from point of origin to final destination. Moreover, the Interstate Commerce Act, as amended, contemplates that the carriers as a whole will, as nearly as possible join into *one national system.*

Under the Act to Regulate Commerce, joint through routes and rates were matters of contract or agreement among carriers, and the Interstate Commerce Commission had no power to compel them to enter into an arrangement for such through routes and rates. However, the passage of the *Hepburn Act* in 1906 gave the Commission authority to establish through routes between points where no "reasonable or satisfactory through route" existed. This limitation upon the authority of the Commission to establish a through route only when no satisfactory route existed was removed in the *Mann-Elkins Act* of 1910 and the Commission was left with full discretion to establish a through route whenever, in the light of all facts and circumstances, it determined that such action was wise, fair, reasonable and equitable.

At the present, the Interstate Commerce Act, Revised makes it the duty of rail carriers to establish through routes and rates and provides the procedure by which the Interstate Commerce Commission may order the establishment of such routes and rates whenever deemed by it to be necessary or desirable in the public interest.

In the opposite vein, however, the Commission's power is no greater or different when dealing with tariffs proposing to cancel an existing through route than in connection with a complaint praying for the establishment thereof. It has no power to prevent cancellation of through routes and joint rates voluntarily established when the circumstances and conditions would not warrant an order to compel such arrangements if not already in effect. In other words, if the proposed cancellation were found "necessary or desirable" it could, under the Act, compel its retention in the public interest. If the "public interest" is not affected, it could not order its retention to any greater extent than it could order its publication in the first instance.

Having established a through route and, therefore, the route being one, a charge for a service over it is a charge for a *single* service, all the terms of which must be fixed at one and the same time; that is, at the time the initial carrier enters into the engagement for the service. The rate is either a *joint through rate* made by arrangement of the parties to the through route, or it is a *combination through rate* consisting of the separately established rates, fares and charges applied to the through transportation. This sum, however, is a *single* rate for a single service, and a contract for through transportation is a contract for transportation at the through rate, whether jointly or separately established, in force at the time the shipment is billed.

When a through local or joint rate has been established via a through route the Commission holds that such local or joint rate is the *only* legal rate for through transportation even though some combination is lower. (Rule 55 (a) of Tariff Circular No. 20.)

In the absence of a published local or joint through rate *via a through route,* the lowest aggregate of intermediate rates applicable via the route of movement applies. If, however, *no through route* has been formed, each separate rate, under a separate bill of lading, must be assessed as of the date each separate bill of lading is signed.

Duty of Carriers to Establish and Operate Through Routes

Section 10703 of P.L. 95-473 provides for the establishment of through routes by carriers subject to the jurisdiction of the Interstate Commerce Commission. In some cases the statute makes it mandatory, such as rail carriers with other rail carriers, and in other cases, such as motor common carriers of property with other such carriers, it is optional.

Similar provisions were contained in the former Interstate Commerce Act and had for their objective the relieving of the shipping public from one of the most vicious evils that the rapid growth and diversified management of our early transportation systems had fostered—the lack of a coordinated system of rates, facilities and practices when the movement of a shipment embraced two or more lines of railway. Their passage marked the beginning of a standardization of service and a merger of the various railroads of the country into a unified system for the receipt and movement of shipments by connecting lines of railroad under through bills of lading for continuous carriage.

In touching upon this feature, the Commission, in the case of *Chamber of Commerce of the State of New York, et al, v. N.Y.C. & H.R. Railroad Company, et al,* 24 I.C.C. 55, 76, stated:

> "The theory of the law is that carriers shall establish and maintain through routes and joint rates so that there may be the freest movement of traffic without the necessity of reshipment."

Under the provisions of the Interstate Commerce Act the railroads were called upon to so unite themselves that they should constitute one national system. They must establish through routes, keep these routes open and in operation, furnish necessary facilities for transportation and make reasonable, and proper rules and regulations with respect to their operation. In other words, each connecting carrier, in the discharge of its duties to the public, owes to shippers of freight in its possession destined to points on or routed over the line of another railroad, the duty to deliver to the connecting line for further transportation; and each carrier is correspondingly bound to receive and carry such freight (*New York Central R. Co. v. Tal., Long Isl. R. C.,* 288 U.S. 239).

As the theory of the law is to combine the railroads of the country into a closely knit and perfectly coordinated system of transportation, the duty of the initial carrier to furnish equipment for a shipment which moves onto other lines is universally recognized, and in cases where that is deemed impracticable or unwise, the carriers, under a through bill of lading, assume to bear the burden of transferring the freight from the equipment of one line to that of the other at the junction point.

In this connection, the Commission, in *Missouri & Illinois Coal Company v. Illinois Central Railroad Company,* 22 I.C.C. 39, 44, said:

> "There can be little doubt as to the duty of the carriers under the present act. The commerce of the country is regarded as national, not local, and the railroads are required to serve the routes which they have established,

or which they may have been required to establish, in connection with other carriers, without respect to the fact that this may carry their shipment beyond their own lines."

While it is the duty of railroads to serve the routes which they have established, or which they may have been required to establish, it must be remembered, there is nothing in the statute which requires a railroad to establish through routes via all possible junctions or by way of all railroads or water lines with which it might connect, or between all points or areas which might be reached through those connections. A carrier has the initial duty and right to select through routes over which it holds out to transport through traffic and participate in through rates. (*Routing Restrictions over Seatrain Lines, Inc.*, 296 I.C.C. 767, 774.)

In Section 216, paragraph (a), *Part II* of the *Interstate Commerce Act*, the following provision was made:

"It shall be the duty of every common carrier of passengers by motor vehicle to establish reasonable through routes with other such common carriers ***; to establish, observe and enforce just and reasonable individual and joint rates, fares and charges, and just and reasonable regulations and practices relating thereto. *** "

Section 216, paragraph (c), provided as follows:

"Common carriers of property by motor vehicle may establish reasonable through routes and joint rates, charges and classifications with other such carriers or with common carriers by railroad and/or express and/or water; and common carriers of passengers by motor vehicles may establish reasonable through routes and joint rates, fares or charges with common carriers by railroad and/or water."

It should be particularly noted that, under Part II of the Interstate Commerce Act, the duty of carriers subject to this Part of establishing through routes with other motor carriers was placed upon common carriers of *passengers* only, with no mandatory requirements as between such carriers and common carriers of passengers under Parts I or III. Insofar as common carriers of *property* via motor truck are concerned, it provided that such carriers *may* provide such through routes with other motor common carriers or with common carriers by railroad, express and/or water. In other words, there was nothing mandatory applicable to carriers of property under Part II, the matter being left to the voluntary action and discretion of the motor carriers.

While Section 216(c) did not specifically prohibit joint rates or arrangements for through carriage between common and contract motor carriers, the fact that it expressly made provision for through routes and joint rates only between common carriers had the force and effect of prohibiting through routes and joint rates or arrangements for carriage between common and contract carriers. (*Chicago and Wis. Points Proportional Rates,* 17 M.C.C. 573, 577.)

In connection with water carriers, Section 305(b), *Part III* of the *Interstate Commerce Act,* provided that it shall be the duty of common carriers subject to this Part:

> "To establish reasonable through routes with other such carriers and with common carriers by railroad, for the transportation of *persons or property* ***. Common carriers by water *may* establish reasonable through routes *** with common carriers by motor vehicle."

It will be noted that Section 305(b) made the establishment of through routes with common carriers by railroad mandatory. Through routes with common carriers by motor vehicle *may* be established but there was no requirement that this be done. See also Section 216(c).

The increasing use of "containers" permitting through intermodal transportation between points in the United States and points in foreign countries without the rehandling of the freight at points of interchange between carriers has resulted in the establishment of many through routes and joint rates. It has, however, also created many problems with regard to the publication of through joint rates and the jurisdiction of the regulatory agencies. This particular phase will be discussed in greater detail in the succeeding chapter under the heading of joint through rates.

Through Routes As Seen From the Legal Viewpoint

Section 10703 of Public Law 95-473 makes it the duty of carriers subject to its provisions to establish through routes and rates applicable thereto. Furthermore, Section 10761 provides that such carriers "shall provide that transportation or service only if the rate for the transportation or service is contained in a tariff that is in effect."

With regard to the rate or rates applicable over a through route Section 6(1) of the Interstate Commerce Act provided:

> "If no joint through rate over the through route has been established, the several carriers in such through route shall file, print and keep open to public inspection as aforesaid, the separately established rates, fares and charges applied to the through transportation."

Similar provisions were also contained in Sections 217(a) and 306(a). However, in the revised Act, Public Law 95-473 the above specific language was omitted as unnecessary since the result would be the same under the broad provisions of Section 10762.

The reasons for these requirements of the statute are at least two: (1) the policy of the law that carriers otherwise not subject to the statute shall be, when participating in interstate business, subject to its restrictions and requirements, and be required to publish their tariffs and file them with the Commission, and (2) the policy of the law that every route and every service shall have a published charge equally known and equally available to all patrons of the carriers.

As to the first feature, the Commission, in the case of *Baer Bros. Mercantile Company v. Mo. Pac. Railway Company and D. & R.G. Railroad Company,* 13 I.C.C. 329, stated:

> "A railroad company whose road lies entirely within the limits of a single state becomes subject to the Act to regulate commerce by participating in a through movement of traffic from a point in another state to a point in the state within which it is located, although its own service is performed entirely within the latter state."

Also, in *Beloff Contract Carrier Operation,* 1 M.C.C 797, held:

> "An interstate line haul service by a motor carrier under contract with the shipper, performed in continuance of and to complete a prior interstate rail movement is an interstate operation" as the term is defined in the Act, and cannot lawfully be carried on without an appropriate permit."

Also, in *Jackson Common Carrier Application,* 19 M.C.C 199, the Commission made the same findings relative to a common carrier operation lying entirely within the confines of a single state as being nevertheless engaged in interstate commerce on traffic destined beyond the state lines.

In *Hudson River Day Line Com. and Contr. Car. Applic.,* 250 I.C.C. 396; *New England S.S. Co. Com. Car. Applic.,* 250 I.C.C. 184, *Red Collar Line, Inc., Com. Car. Applic.,* 250 I.C.C. 785; it was held that water carriers were engaged in interstate transportation although their operations were physically intrastate, when they transported passengers or property, or performed towage service, on through tickets or under joint rate or other arrangements with carriers operating to points outside the State.

In amplification of the second feature, we find the following expression of the Commission *In the Matter of Through Routes and Through Rates,* 12 I.C.C. 163, 166, 167:

> "Carriers arranging for a through route and also for a joint rate must give notice to the world of such arrangement by publication. Carriers for-

ming through routes without joint rates, however, need publish and file only 'the separately established rates, fares, and charges applied to the through transportation.' As a matter of fact, most, if not all, carriers subject to the Act use their local rates as through rates when no joint rate is established, neither publishing nor filing such 'separately established rates, fares, and charges, applied to the through transportation' otherwise than in the tariffs of such local rates."

Here, then, we have the express recognition in the statute of the possibility of a through route being in effect over the line of two or more carriers despite the fact that concrete evidence of its existence may not be given by the publishing of a joint through rate (a one figure rate) in a single tariff to which all carriers making up such through route are parties. In other words, while carriers may get together and combine their services into a single system between certain points, thus making a through route, it is not necessary, nor is it required by the statute, that they also combine their tariffs applying over such through route into a joint issue. They may imply or indicate to the shipping public in certain ways that such through route actually does exist.

Facts Determining Existence of a Through Route

The question, therefore, naturally arises as to the test the shipper may apply or the nature of the implication that will determine whether a through route is actually in existence when no joint through rate has been published to cover.

In passing upon this feature, the Commission, *In the Matter of Through Routes and Through Rates, supra,* stated:

"The existence of a through route is to be determined by the incidents and circumstances of the shipment, such as the billing, the transfer from one carrier to another, the collection and division of transportation charges, or the use of a proportional rate to or from junction points or basing points. These incidents named are not to be regarded as exclusive of others which may tend to establish a carrier's course of business with respect to through shipments."

Also, in the same report we find the following:

"The conclusion to which we are compelled is that where through billing is given by the originating carrier and is recognized by all connecting carriers to destination, there is in existence a through route over which a through rate applies, which through rate is ascertainable from the tariffs of the participating carriers at the date of shipment."

In *Swift and Company v. P.R.R. Co. et al,* 214 I.C.C. 464, 466, and in *Beaman Elevator Company v. C. & N.W. Ry. Co.* 155 I.C.C. 313, 317, the following concrete finding was made to the circumstances that may prove the existence of a through route:

> "Therefore it is settled that, whatever other facts or incidents of a shipment may serve to prove the existence of a through route, a through bill of lading is, as to carriers recognizing it, conclusive evidence of the existence of such through route."

If, therefore, the initial line issues a through bill of lading on a shipment and such through bill of lading is recognized as such by the successive carriers over whose lines the shipment passes to point of final destination, the shipper may be satisfied that there is a through route in existence, even though there is no joint through rate published to cover such movement.

The Supreme Court has held that combination rates necessarily reflect agreement, express or implied, between connecting carriers to establish a through route for continuous carriage from origin on one to destination on the other. Each proportional necessarily is a part of the through rate and is capable of use only as such. (*Great Northern Ry. Co. v. Sullivan,* 294 U.S. 458, 460.)

In the event there is no through route applying from point of origin to the destination desired by the shipper, the initial carrier would so indicate on the bill of lading by making a notation thereon to the effect that the billing it was issuing would apply only to a specified junction point, and not to a final destination.

The same general enunciations concerning through routes and/or through rates apply with equal force to carriers by motor vehicle.

In *Vermont Transit Co., Inc.,* 11 M.C.C. 307, the Commission said with reference to arrangements between connecting lines:

> "A motor carrier and its connection have an undoubted right to establish joint rates for the movement of traffic under through bill of lading and to divide such joint rates justly, reasonably and equitably. In the absence of joint rates, traffic may be moved under through bill of lading, each carrier charging its own local or proportional rate for its portion of the through haul. *** "

And, further, the Commission stated in *Rush Common Carrier Application,* 17 M.C.C. 661, that:

> "In determining whether in particular instances a joint arrangement for continuous carriage exists, the Commission may appropriately consider the holdings of the Supreme Court and its own decisions in similar questions under Part I of the Interstate Commerce Act."

Therefore, in construing the existence of a through route, etc., we find a definite relationship and application of precedents in determining joint arrangements for continuous carriage by motor truck under decisions applicable to such arrangements via rail.

In the final analysis it must be remembered that a shipper can make a through shipment where no through route has been established provided the necessary interchange facilities exist for, while nothing in the statute requires a carrier to establish through routes and joint rates via all possible connections, it may not, except under extra-ordinary conditions, lawfully refuse to transport a shipment from or to any point on its lines which it holds out to serve, including an established interchange with a connecting line. (*High Point Chamber of Commerce v. Southern Ry. Co. et al.,* 314 I.C.C. 683.)

Cancellation of Established Through Routes

As we have seen, Section 10703(a)(4)(A) permits common carriers by motor vehicle to establish through routes with other such common carriers. Once established, however, does the carrier have the right to cancel such through routes with respect to certain specified commodities or in connection with certain carriers?

In 1966 Associated Truck Lines cancelled all through routes and joint arrangements on shipments of furniture between Central States Territory and Southern Territory. The National Furniture Traffic Conference filed a complaint with the Commission challenging the lawfulness of such cancellation. The Commission in *National Furniture Traffic Conference v. Associated Truck Lines,* 332 I.C.C. 802, ruled that the cancellation of the through routes was unlawful stating:

> "This Commission has previously found that under section 216(c) it has no authority to require the establishment or maintenance of through routes and joint rates. However, as noted, there is nothing in our prior reports which prevents us from ending unreasonable practices of the kind found to exist here.
>
> "Furthermore, Associated's selective holding out clearly places furniture in a disadvantaged position relative to other general commodities. We find such disadvantage to be undue and unreasonable, and, hence unlawful under section 216(d).
>
> " *** Accordingly we find that the selective cancellation of through route arrangements on specified classes of commodities will adversely affect the public interest and section 216(c) does not foreclose us from observing our plain duty to preclude this unreasonable practice."

The decision of the Commission was upheld by the Federal District Court, 304 F. Supp. 1094 (1969) and affirmed by the United States Supreme Court, 297 U.S. 42 (1970).

In another, and later proceeding(1971), the Commission found the cancellation by McLean Trucking of through routes, joint rates and interchange arrangements with Manning Motor Express unjust and unreasonable. *Interchange between McLean Trucking and Manning Motor Express,* 340 I.C.C. 38. An action was filed in the Federal District Court seeking to set aside the Commission's order. The court in upholding the Commission's decison, said:

> "In its opening brief as plaintiff, McLean does not mention Section 216(g). McLean confines its argument to considerations affecting Section 216(c); it argues that since the Commission has no power to require establishment of joint rates, it has no power to require their continuation, and in argument counsel for McLean suggest that 216(c) is a bar to any Commission control over discontinuance of joint rates.
>
> "McLean's conclusion, though not required, is tenable, so long as 216(g) is ignored.
>
> "However, Section 216(g) will not be ignored. It says in plain English that when a carrier proposes to make a change in a fare or in a 'rule, regulation or practice,' the burden of proof is upon the carrier to show that 'the proposed changed rate, fare, charge, classification, rule, regulation, or practice is just and reasonable.' "
>
> "McLean has offered no proof to sustain its burden of proof under Section 216(g), but relies essentially upon its theoretical assertion that what it can not be required to commence, it can not be required to continue.
>
> "The Commission correctly determined that the carrier has not sustained its burden under section 216(g) to prove that the proposed new tariffs and procedures were 'just and reasonable' and correctly denied the carrier authority to discontinue the joint rates with Manning Motor Express. *McLean Trucking v. U.S.,* 346 F. Supp. 349 (1972), affirmed by the United States Supreme Court (1973).' "

Rate Applying Over a Through Route

Many of the controversies between shippers and carriers arising out of the question of existence of a through route are brought about by the rates of one or more of the carriers involved undergoing a change while the shipment is in course of transportation. The rate to apply in cases of this kind can best be determined by following the reasoning of the Commission as expressed in, *In the Matter of Through Routes and Through Rates,* previously referred to, as follows:

"Upon the request of various shippers and railroads a hearing was held at Chicago *** upon the matter of through rates which should govern shipments, over more than one railroad where no joint rate has been made. A typical case presented was that arising out of the claim of the Crane Company, of Chicago, against the Oregon Short Line. This case involved the rate to be applied to a shipment originating at a point east of Chicago and moving through Chicago and Omaha to Utah. A through bill of lading was issued by the originating carrier from the point of origin to the point of destination in Utah. No joint rate had been made by the carriers who participated in the through movement under the through bill of lading. The shipment moved upon the rate to Chicago plus the proportional rate from Chicago to the Utah destination. After this shipment had been billed, and while it was in transit, but before its arrival at Chicago, the proportional from Chicago to Utah points was reduced, and this reduction had become effective by proper publication and filing. The carriers applied to the shipment from point of origin to destination the rates in legal effect at the time of billing, refusing to give the shipment the benefit of the reduction in the rate beyond Chicago which became effective after the service on the through bill of lading had been begun. The shipper, the Crane Company, claimed the benefit of the reduction. It was testified by both shippers and carriers that much controversy is arising in such cases as the above. It was agreed that if shipments are to enjoy reduction in rates made while in transit, they must be subject to increases made under like conditions.

"The problem presented may be thus stated:

"In case a shipment has been made over two or more railroads which have not, as to the journey the shipment is to take, filed with the Commission a notice of through route and joint rate as required by Section 6 of the act to regulate commerce, does the shipment take the sum of the local rates of the various lines over which it is moved as such locals may be established at the time it is received by the initial carrier, or is it subject to changes in locals which may be made before the shipment reaches the lines making such changes? That is to say, a shipment is made over the A line to a connection with the B line, and thence over the B line to destination. The B line changes its local after the shipment is billed at point of origin, but before it has passed beyond the A line. Is the shipment subject to such change in rate of the B line, or does it move under rates in effect at the time it began its journey?

"A careful consideration of all the factors entering into this problem shows that in the last analysis the answer must depend upon a question of fact, this question being, Have the carriers over whose lines the shipment is to move made an arrangement, express or implied, for a through route? If a through route has been so formed, then the rate charged must be a through rate, and the shipment will move upon the rate existing at the time it is billed. If, however, no through route has been formed, then the shipment will move, not upon one through journey, but upon a succession of journeys, and will be subject to any change in rates made by any carrier into whose possession the shipment has not been received."

Upon analyzing this expression of the Commission, and more especially the final paragraph, it will be observed that where through billing is given by the originating carrier and is recognized by all connecting carriers to destination, or other incidents and circumstances so indicate, there is in existence a through route over which a through rate applies, which through rate is ascertainable from the tariffs of the participating carriers in effect at the time the shipment begins to move. Further, that when such through rate is made up of the sum of the locals, *the locals apply as of the date of shipment* and not as of the date the shipment is received by each individual line. Therefore, any increase in the through rate so made up after the date of shipment, or any decrease therein, is not applicable to such through shipments.

After a through route has once been established between two or more carriers, such through route is a new unit—a single line from point of origin to destination made by an agreement, express or implied, between such carriers. The rate applying over such through route must likewise be a unit even though it may be made up of several factors, which unit must be preserved intact from the date the shipment begins its journey over the through route.

Authority of Commission to Prescribe Through Routes

In Section 10705 of P.L. 95-473 it is provided that:

"The Interstate Commerce Commission may, and shall when it considers it desirable in the public interest, prescribe through routes, joint classifications, joint rates (including maximum or minimum rates or both), the division of joint rates, and the conditions under which those routes must be operated, for a common carrier providing transportation subject to the jurisdiction of the Commission under subcapter I, II (except a motor common carrier of property), or III of chapter 105 of this title. When one of the carriers on a through route is a water carrier, the Commission shall prescribe a differential between an all-rail rate and a joint rate related to the water carrier if the differential is justified."

In *Great Western Packers Exp. Inc., v. U.S.* 246 F. Supp., 15, the court ruled:

"Under 216(c) motor common carriers of property may estabish through routes and joint rates, but no statute requires them, or authorizes the Commission to require them, to do so. Therefore, carrier was free to discontinue its participation, with plaintiff and a third carrier, in tariff establishing through routes and joint rates for three line transportation of meats and packing house products. After it did so the three line joint rate was no longer effective, since the through operation was eliminated by its action of withdrawing from participation."

Prior to 1910 the power of the Commission to establish through routes was limited to instances in which no satisfactory through routes existed. However, under the Interstate Commerce Act, as amended, this limitation was eliminated, and in the event a carrier refused, under Parts I or III, to establish a through route it was within the discretion of the Commission to say whether the public interest makes such through route essential, irrespective of the fact that another suitable through route may be in existence between the points in controversy.

It is understood, of course, that in giving the Commission authority to prescribe through routes and joint rates, it was not intended to require it to prescribe such through routes and joint rates whenever requested to do so, without regard to the peculiar circumstances of each case. The necessity for a through route must be apparent or otherwise the Commission will not order its establishment. In this connection, two things are certain, as stated by the Supreme Court in *W.P. Brown & Sons Lbr. Co. v. Louisville & N.R. Co.,* 299 U.S. 393, 399:

"If the route to which alone the joint through rate applies is deemed inadequate, there is power in the Commission to establish other through routes with joint rates, and if rates on combination routes are excessive or discriminatory, the Commission has power to reduce them."

Right of a Carrier to the Long Haul In Prescribed Through Route

After delegating to the Commission the authority to prescribe through routes and joint rates, Section 10705(a)(2) of P.L. 95-473 restricts this authority as follows:

"The Commission may require a rail carrier to include in a through route substantially less than the entire length of its railroad and any intermediate railroad operated with it under common management or control if that intermediate railroad lies between the terminals of the through route only when—

(A) required under section 10741-10744 or 11103 of this title;

(B) one of the carriers is a water carrier;

(C) inclusion of those lines would make the through route

unreasonably long when compared with a practicable alternative through route that could be established; or

(D) the Commission decides that the proposed through route is needed to provide adequate, and more efficient or economic, transportation."

In interpreting Section 15(4) of the Interstate Commerce Act the Commission, in the case of the *Waverly Oil Works Company v. P. R.R. Co., et al,* 28 I.C.C. 621, 629 stated:

"This power to establish a joint rate must, however, be exercised under the limitations imposed upon this Commission in the fixing of joint rates, and these limitations should be kept clearly in mind. Originally the statute contained no provision permitting the establishment of a joint rate between two carriers by public authority, and the courts held that, notwithstanding the provision of the third section, railroads were free to select their connections and to make such arrangements for the handling of through business with those connections as they saw fit. To make such arrangements with one railroad and exclude another was not an undue discrimination.

"Later the Act was amended so as to give this Commission authority to establish a through route between points where no 'reasonable and satisfactory route' already existed. If a railroad had provided one satisfactory route for the transaction of business, either all the way by its own line or by its own line in connection with some other line, this Commission had no authority to establish an additional route. The manifest intent of this provision was to permit a railroad to handle by such route as it saw fit traffic which it could obtain, provided it offered a satisfactory route and therefore protected the public interest.

"Still later this provision for joint rates was changed so as to allow the establishment between two points of an indefinite number of routes, but it was now provided that in the establishing of a through route no railroad should be required to haul traffic over less than the entire length of its line unless such route was unduly circuitous. The purpose of Congress here again is plain. If a railroad has traffic in its possession, it shall be allowed to handle it by its own line as far as it can unless the public interest will suffer thereby.

"A question arises in the practical application of this statute which will be most easily understood from the diagram on page 51.

"Railroad No. 1 extends from A to X and from A to Y; railroad No. 2 from B to Y. If, now, a through route is to be established between A and B shall it make via the junction at Y or via the junction at X? The statute does not in terms answer this question, but the manifest intent of Congress was to provide that a shipment from A should be routed via junction X giving Railroad No. 1, which initiates and has possession of the traffic, the

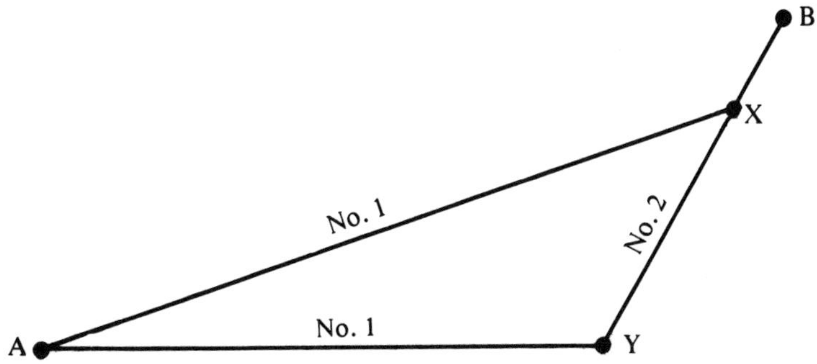

long haul, while a shipment originating at B should be routed via junction Y, thus giving to railroad No. 2, which has possession of the traffic, its long haul."

This provision of the law, as well as its interpretation by the Commission indicates that while a carrier is ordinarily protected in its long haul on the traffic it originates (including the lines of its subsidiaries), such right is subservient to the welfare of the shipping public. A carrier may, therefore, be required by the Commission to forego its long haul in the establishment of a through route in the event the inclusion in such through route of all or substantially all of its line or lines between the points in question would result in an unreasonably long route as compared with another practicable through route which could otherwise be established, or would trespass in any other manner upon the rights of the shipping public. This is supported by the Commission's opinion in *City of Sheboygan v. Chicago N.W. Ry. Co.* 227 I.C.C. 472, 478, wherein it stated:

> "In the absence of undue prejudice or an unduly circuitious route, the Commission cannot require a carrier to turn traffic over to a connecting line when to do so would require it to participate in a through route embracing substantially less than the entire length of its line between the points in question."

The law imposes certain obligations upon a carrier when it joins the connecting carriers in establishing through routes. It is not unreasonable for such carriers to refuse to assume those obligations over all routes, and it cannot be required under the law to assume those obligations over a particular route merely because a shipper or connecting carrier selects it.

There is nothing in the statute, however, that requires a railroad to establish through routes via all possible junctions or between all points or areas which might be reached through those connections. On the contrary, the law is definite that a carrier has the initial duty and the right to select the routes over which it holds out to transport through traffic and participate in through rates.

CHAPTER 4

JOINT THROUGH RATES

General

In this and several subsequent chapters the legal nature of a freight rate will be dealt with based primarily upon the former Interstate Commerce Act and the present statute, P.L. 95-473 as interpreted by the Interstate Commerce Commission and the courts. There is, of course, a similarity between the State and Federal regulatory statutes. State laws generally conform to the Federal law insofar as the regulation of rates is concerned in that State laws also require all rates to be reasonable and nondiscriminatory.

The subject of joint rates, which this chapter covers, and the legal principles which underlie their formation, i.e., the joining of two or more lines into a single system for the preservation of the through rate, is of utmost importance. The policy of the law and the convenience of business favor the making of joint rates, and the more completely the systems of the country can be considered and treated as a unit, the greater will be the benefit of the services to the public. The plain, unmistakable intent and requirement of the law is, therefore, that such traffic shall be subject to just and reasonable through rates.

When it is shown that a through rate is a great convenience to shippers, adds materially to available transportation facilities, and is therefore a substantial benefit to commerce, it must be held to satisfy the test of public necessity, as the establishment of joint rates depends upon a showing that they *are* necessary in the public interest.

A through rate may be described as an arrangement, expressed or implied, between connecting carriers, for continuous carriage from a point on the line of one carrier to a point on the line of another carrier. The through rate need not be a joint rate but may be a combination of locals or proportionals. Therefore, while a joint rate is not essential to show the existence of a through route, its establishment presupposes through routes. Likewise a combination of proportional rates necessarily reflects an agreement to establish a through route. Where joint rates or combinations of proportionals do not exist, and traffic moves under combinations of local rates or of local and proportional rates, the facts in each case determine whether an agreement to move traffic through to ultimate destination in continuous carriage may be implied.

The carrier may initiate and publish any through rate or combination of rates, or any through route it so desires as there is no presumption of wrong arising from a change in a rate or route by a carrier. The presumption of honest intent and right conduct attends the action of carriers as well as it attends the action of other corporations or individuals in their transactions in life. Undoubtedly, when rates are changed, the carrier making the change must, when properly called upon, be able to give a good reason therefor; but the mere fact that a rate is raised or lowered carries with it no presumption that it was not rightfully done. The Commission and the courts have always recognized the right of carriers to adapt their rates and practices to the needs of localities and industries which they serve, so as to promote the welfare of shippers and carriers alike, within the limits of reasonableness and without undue prejudice and unjust discrimination.

Whenever the carrier transgresses such limitations or refuses to establish a through rate, the power of the Interstate Commerce Commission may be invoked, as the issue of necessity and convenience affecting the public interest must be satisfied.

When through routes or rates have been established by voluntary action of the carrier, the request of the shipper, or by order of the Commission, carriers, party to such agreement, are jointly and severally responsible for that rate, and those carriers who actually participate in the transportation under a joint rate are jointly and severally liable for the reasonableness or unreasonableness of it. But where through rates are made up of separate locals of individual carriers, none of whom control any factor except that maintained by it, or contribute to injury except to extent of unreasonableness in its particular rates, to find in such instances that carriers are jointly and severally liable, would be to require one or more carriers to bear injury done by another participating carrier or carriers where former had no control or voice over actions or rates of the latter.

Carriers entering into such joint arrangements are easily ascertainable to those interested in using such rate. The Commission has provided in its Tariff Circulars that the issuing carrier indicate in the tariff the name of each participating carrier. It also requires the form and number of the power of attorney, or certificate of concurrence, given the issuing carrier. However, the reference to such forms and numbers may be omitted provided this information is furnished to the Commission in acceptable form.

Through Rate Defined

Briefly speaking, a through rate is the rate applicable over a through route from point of origin of a shipment to its destination. It may be: (1) a local rate, (2) a joint rate, or (3) a combination through rate, i.e., a combination of separately established rates applicable to through business over a through route which does not enjoy a joint rate. This was defined by the Commission in *Scoular-Bishop Grain Co. v. M. & St.L. R.R. Co.*, 200 ICC 665, as follows:

> "The term 'through rate' . . . means the joint through rate where one is available, and in the absence of an available joint through rate, the lowest combination in effect over route of movement is the 'through rate.' "

The Commission, in using the power conferred upon it to prescribe regulations as to the form and manner that tariffs of motor carriers shall be filed with it, defines a "through rate" in its Tariff Circular as:

> "The term 'through rate' as used herein, means the total rate from point of origin to destination. It may be a local rate, a joint rate, or a combination of separately established factors."

A local rate is, strictly interpreted, a through rate, inasmuch as it applies through from point of origin to destination. However, it is seldom referred to in such manner, and need not, therefore, be considered further in connection with our present discussion. This leaves the features of a joint through rate and a combination through rate for analysis.

Joint Through Rate Defined

A joint rate, as stated, is a one-figure rate that applies over the lines of two or more carriers and is made by agreement between such carriers. A joint rate is, therefore, a through rate as to the traffic handled from and to the points between which it applies.

The term "joint rate" has acquired the definite meaning that two or more carriers forming a through route have agreed upon and join in a combined total rate known as a "joint rate" instead of requiring the payment of each carrier's individual rate.

It is this form of through rate, legally referred to as a joint throught rate, that generally though erroneously, is understood to be the only form of through rate. However, this is not the case, as a through rate may also be a combination through rate or a local rate.

This was brought out by the Commission in *Parkersburg Rig & Reel Co. v. B. & O. R.R. Co.*, 225 ICC 581, in the following manner:

> ". . . the expression 'through rates' does not necessarily mean joint through rates. A through rate may be a joint rate or merely an aggregation of separate rates fixed independently by the several carriers forming the through route, as where the through rate is the sum of the locals on the several connecting lines, or it may be the sum of lower rates otherwise separately established by them for through transportation."

Legal Nature of a Joint Rate

The essential element of a joint rate is the fact that there must be a *definite agreement* between the carriers involved to join their lines into a single system. This is evidenced by the following remarks of the Commission, as contained *In the Matter of the Form and Contents of Rate Schedule, and the Authority for Making and Filing Joint Tariffs,* 6 ICC 267, 268:

> "Such tariffs (joint tariffs) necessarily imply an agreement between two or more carriers by virtue of which they offer their united services at the rates therein named. They must have entered into contract relations with each other, by express stipulation or mutual understanding, which impose upon the several parties thereto the obligation to accept, for the transportation proposed by them, the aggregate charge stated in their advertised schedule. . . . If one carrier files a tariff purporting to establish joint rates with other carriers when no agreement thereof—expressed or implied—exists between them, such carrier transcends its authority, misrepresents the facts and misleads the public."

In the case of *The Consolidated Forwarding Company v. The Southern Pacific Company, et al.,* 9 ICC 182, 205, the Commission stated:

> "Every continuous rail line or route authorized by the 6th Section of the Act is of necessity constituted by two or more separate roads uniting by voluntary agreement and fixing joint through rates over the lines thus formed. Such a route is in every instance as definite and specific a physical line as is either of the separate roads which constitute it."

In the making of freight rates it must be remembered that the carriers have the right to initiate their rates and as far as local rates are concerned, the only authorization necessary is the publication and filing of the rates by the carrier over whose line they apply. Where through rates are made by a combination of locals no authorization is necessary, each of the locals having been published by the individual carriers involved. However, in the case of a joint rate, each of the carriers over whose line the traffic moves usually receives a division or percentage of the through rate which is in most cases less than its full local. It is apparent therefor that such a joint rate cannot be established except with the agreement and consent of the carriers involved. A joint rate can never be properly made unless the minds of the carriers have met in full accord and they have agreed to the incidents which go into the making of such a rate.

In this connection, the Commission, in the case of *The Diamond Mills v. Boston & Maine Railroad Company,* 9 ICC 311, stated:

> "At common law, and under the Act to Regulate Commerce as interpreted by the courts, joint through routes and through rates are matters of contract between the connecting carriers."

Also in *Canton R. Co. v. Ann Arbor R. Co.,* 163 ICC 263; and *Southern Roads Co. v. Galveston, H. & S.A. Ry. Co.,* 168 ICC 768, the Commission stated:

> "A joint rate is made, unless required by the Commission in a proper proceeding, by arrangement or agreement between such carriers and evidenced by concurrence or power of attorney."

It is well settled, therefore, that in order to have a joint rate there must be an express agreement or contract between the carriers whose lines constitute parts of the through movement and that if such agreement is lacking there is, in the eyes of the law, no joint rate. A carrier may not, therefore, lawfully publish a joint rate unless it is authorized by agreement with another carrier to do so.

It was in recognition of this principle—that every rate must be the result of a definite agreement between the carriers interested—and that the shipping public should be definitely informed of such contract, that P.L. 95-473 provides in Section 10762(b)(2):

> "A joint tariff filed by a carrier providing transportation subject to the jurisdiction of the Commission under subchapter I of that chapter shall identify the carriers that are parties to it. The carriers that are parties to a joint tariff, other than the carrier filing it, must file a concurrence or acceptance of the tariff with the Commission but are not required to file a copy of the tariff. The Commission may prescribe or approve what constitutes a concurrence or acceptance."

It will be noted that the above is applicable only to subchapter I carriers (railroads) and there is no comparable provision applicable to motor common carriers. However, paragraph (a)(1) of Section 10762 states "The Commission may prescribe other information that motor common carriers shall include in their tariffs." Under this authority the Commission has included a rule in their Tariff Circular requiring the names of the participating carriers to be shown in their tariffs.

Although Section 10762 contains nothing specific requiring water carriers to publish in their tariffs the names of the carriers participating in joint rates the Commission's Tariff Circular does provide that such information shall be shown.

In administering these provisions, the Commission, by means of a system of concurrences, has provided for a carrier definitely indicating its agreement to the joint rate or rates contained in a joint tariff (issued either by another carrier or by an agent acting for several carriers).

Under this arrangement the Commission knows definitely whether a carrier has agreed to the joint tariff and is able to determine to what extent it participates in the rates contained in such tariff.

The Commission's tariff regulations now require that the carrier or joint agent issuing a joint tariff shall, *before issuing it,* have secured a definite and affirmative concurrence of every carrier shown therein as a participant, and shall show in connection with the name of each participating carrier, the form and number of the instrument by authority of which the carrier is made a party to the tariff. References to the forms and numbers of powers of attorney and concurrences may be omitted provided this information is furnished to the Commission, in an acceptable form.

Purposes of Concurrences and Power of Attorney

Briefly, these concurrences convey authority from one carrier to another carrier or agent to publish joint rates. They are designated by the symbols *"FC"* covering freight tariffs of railroads; *"MFXC"* via motor, and *"FX"* when via water carriers. These convey varying degrees of authority, according to the form issued, as shown on page 59.

Shipper Not Bound by Nature of Concurrence

An element of primary importance to be borne in mind in connection with concurrences is that they are designed solely for the use of the Commission in determining whether the publishing carrier or agent has authority to publish rates for the various carriers that are shown as parties to the tariff, as well as the extent of such authorization.

It is quite obvious, therefore, that in applying a tariff according to its terms it is not necessary for a shipper to concern himself with the exact nature of the concurrences of the various carriers shown as participating

Railroads

FORM NO.	TO WHOM ISSUED	AUTHORITY CONFERRED
FC 1	Carrier	Unlimited and covers all freight tariffs on traffic moving from, to, via or at points on the line of the carrier giving the concurrence.
FC 2	Carrier	Limited but affirmatively specified traffic moving from, to, via or at points on the line of carrier giving the concurrence.
FC 3	Carrier	Restricted to freight tariffs applying on traffic moving to points on or via the line of the carrier giving the concurrence but is not otherwise limited.
FC 4	Carrier	Limited but affirmatively specified traffic moving to points on or via the line of the carrier giving the concurrence.
FA 1	Publishing Agent	Power of attorney. Unlimited authority to publish local rates and charges for the carrier issuing the power and to publish joint rates for such carrier and such other carriers as have issued the necessary authority.
FA 2	Publishing Agent	Power of attorney. Limited in any appropriate manner provided the limitations are specifically and unambiguously expressed in the instrument.
FA 3	Carrier	Power of attorney. Limited or unlimited.

Motor Carriers of Property

FORM NO.	AUTHORITY CONFERRED
MFXC 2	Authority to another carrier to publish rates in a particular tariff.
MFXC 3	Authority to another carrier or its agent to publish rates in tariffs issued by such carrier or agent.
MFXA 2	Power of attorney in appointing an individual to act as agent.
MFXA 3	Power of attorney in appointing a corporation to act as agent.

As will be noticed the first two listed are concurrence forms while the last two are power of attorney forms.

Common Carriers by Water

FORM NO.	AUTHORITY CONFERRED
FX 10	Power of attorney in appointing an agent.
FX 11	Authority to publish rates for its account.

therein. It is sufficient if he merely refers to the list of participating carriers and determines that all carriers he is using in his routing are listed. He need not check the individual concurrences to verify whether the rates shown are published in accordance with the particular type of concurrence given by the participating carriers to the publishing agent or carrier. For example, let us assume that the Consolidated Rail Corp. is shown as a party to a tariff issued by the Norfolk and Western under an FC 4 authority. As previously shown FC 4 grants authority to publish rates to or via but not from points on the line of the concurring carrier. Assuming further that through error, points of origin on the Consolidated Rail Corp. were shown in the tariff, thus making that carrier a party thereto from points on its line. The shipper would still be protected on the through rates published in such tariff when using the Consolidated Rail Corp. as an originating line. The fact that the FC 4 concurrence, which the Consolidated Rail Corp. had given to The Norfolk and Western did not authorize it to publish rates from points on the Consolidated Rail Corp. would not control.

This principle was originally stated by the Commission in the case of *Healy & Towle v. C. & N.W. Ry. Co.*, 43 ICC 83, and later affirmed in *National-American Wholesale Lumber Association, Inc. v. A.C.L. R.R. Co., et al.*, 120 ICC 665. Following is the pertinent portion of the Commission's decision in the first-mentioned case:

> "Concurrence sheets are not posted in the same manner as are tariffs, and no opportunity is afforded the general public to ascertain whether or not the terms of the concurrence limit the application of the tariff insofar as the participating road is concerned. The tariff of the North Western offered to the public a rate of $82.50 per car on horses from South Omaha to Wausau over the route of movement, and as to the shipments from that point the $82.50 rate must be protected."

It is understood, of course, that the foregoing deals exclusively with the particular kind of authorization that may be incorporated in the various kinds of concurrences. That is to say, for practical purposes, the shipper, in checking rates, need not concern himself with the nature of concurrence forms FC, FA, MFXC, etc., which may be shown opposite the individual carriers shown as parties to the tariff. The tariff itself, in its application, must reflect the limitations imposed on the publishing agent or carrier by such concurrences. As indicated earlier, this information may be omitted provided it is furnished to the Commission in the proper form.

In *South Atlantic Traffic Bureau v. Seaboard Air Line R.R.*, 300 ICC 313, the Commission said:

"It is not important that neither the owner nor lessee of the wharf participated in the publication of the wharfage charge. In such circumstances the carrier or carriers responsible for the publication of the charges must protect them."

Carrier Bound Only by Specific Concurrence in Joint Tariff

While the tariff rules of the Commission require that there must be a definite concurrence by all interested carriers for every joint rate offered the shipping public, it is easily recognized that there is no way of preventing one carrier or agent, through error or otherwise, from naming another carrier as a party to a joint tariff, even though such other carrier had not authorized it to do so. As the carrier so named had not assented to its lines being included in the through route or routes created by such a tariff, it naturally follows that the most essential element of a joint rate (agreement by all carriers making up the through route) would be missing, and that there could, therefore, be no lawful joint rate in effect. Proceeding from these premises, the question then arises as to the standing that a tariff of this nature, and the rates it contains, would have in connection with shipments moved on the strength of its representations.

The Commission, in passing upon this feature, stated in Rule 52 of Tariff Circular 20 as follows:

"(a) The Commission's tariff regulations require that the carrier or agent that issues a joint tariff shall, before issuing same, have secured definite and affirmative authority from every carrier shown therein as a participant, and shall show in connection with the name of each participating carrier the form and number of the instrument by authority of which the carrier is made a party to the tariff.

"A carrier has no means of preventing another carrier from naming it as party to a joint tariff without proper authority so to do, or of preventing another carrier from exceeding the authority conferred by a limited concurrence. It cannot, however, be bound by such unauthorized act, and it is its obvious duty to refuse to recognize or apply any such unlawful issue. It should also at once call attention of the Commission and of the one that issued the tariff to such erroneous action.

"(b) If one or more carriers are, without proper authority, so shown as participating in any tariff and other carriers are lawfully shown as parties thereto, the use of the publication is unlawful as to the carriers that are named as parties thereto without proper authority and lawful as to those that are parties to it under proper authority. The carrier over whose line shipments are sent under a joint tariff is bound by the terms of that tariff if it has lawfully concurred therein, and, if it has not lawfully concurred therein, may not accept earnings in accordance therewith, but must demand for the service performed its lawful earnings according to its lawful tariffs.

"(c) Responsibility and liability for the unlawful incorporation of any carrier in a tariff, or for exceeding the authority conferred by a limited concurrence, will rest wholly upon the carrier that issued the tariff."

In promulgating Tariff Circular M.F. No. 5, the Commission provided that:

"When authority is given an agent to publish rates for a carrier participating under authority of a concurrence to another carrier for whom such agent acts, care must be exercised that rates published for the concurring carrier do not exceed the scope of authority given."

In other words, the agent or carrier publishing rates for the account of another is strictly bound by the wishes of the latter, and in the event no authority whatever is obtained, rates cannot lawfully be published by such agent.

In *Mid-Western Motor Freight Tariff Bureau, Inc. v. Eicholz,* 4 MCC 755, the Commission stated:

"A motor carrier named as a participating carrier in a rate tariff which is the subject of a complaint before the Commission is not a proper party defendant if neither the tariff or any supplement thereto indicates that such carrier gave its concurrence to the publishing line."

Here, then, is a reiteration by the Commission applying to motor carriers the same principles found applicable to the rail carriers in many, many instances. In other words, rates in such tariffs need not and cannot be recognized by the carrier whose authority and concurrence of such participation has not been obtained.

As the Commission so clearly brings out in such instances, there cannot, lawfully, be a joint rate in a tariff unless there is a specific agreement, evidenced by concurrence, between the carriers interested. This does not mean, however, that the shipper will not receive the benefit of such unlawfully published joint rate. In *American Cast Iron Pipe Co. v. A.B. & C. R.R. Co.,* 201 ICC 454, the Commission, in dealing with such a situation, said:

"The rate of $14.00 was published and the fact that the delivering line did not participate therein did not nullify its application. A shipper is entitled to the published rate even though, as here, it was published without authority from one of the necessary carriers. Under such circumstances, the carriers responsible for the publication of the rate of $14.00 must protect it."

Also, in the case of *Du Pont de Nemours Powder Co. v. Wabash Railroad,* 33 ICC 507, which involved a similar situation, it was held that the shipper should get the benefit of the joint through rate because he had relied upon the publishing carrier's representation that the joint through rate was in effect. It was held further that the other carriers were entitled to the full local rates and the carrier publishing the unauthorized joint rate received only the remainder of the rate after deducting such full locals.

The finding of the Commission, that the shipper should be protected in cases of this nature, is based not upon any forced construction as to the legality of a joint rate published in this manner, but to give the shipper the benefit of the doubt. It is, rather, based upon the damage which has been suffered by the shipper in relying upon the unlawful joint rate published by the carrier at fault.

With the growth of T.O.F.C. service and the substitution of rail for motor carrier service it became necessary to clarify the situation with regard to participation in joint intermodal tariffs and on the subject of concurrences the Commission, in *Movement of Highway Trailers by Rail,* 293 ICC 93, said:

> "In substitution of rail for motor service, motor rate tariffs appropriately should contain concurrences of the rail carriers, in conformity with Tariff Circular 20, which requires that tariffs of motor-rail joint rates 'shall contain . . . corporate names of carriers participating under concurrences.' "

Divisions of Joint Through Rates

In the early days of railroading, joint through rates were unknown and the necessity of arranging divisions of revenue did not confront the carriers at that time. When traffic moved over more than one carrier, each line charged its local rate for the distance it hauled the shipment. As time went on, however, joint through rates were established applying over the lines of two or more carriers and the revenues derived from such rates had to be divided.

Paragraph 6 of Section 15 was added to the Interstate Commerce Act by the Transportation Act of 1920, giving the Commission the authority to prescribe just, reasonable and equitable divisions of joint rates, fares and charges. This paragraph has been superseded by Section 10705(b) of P.L.95-473 which provides in part that:

> When prescribing the divisions of joint rates of a rail or water carrier under this subsection, the Commission shall consider—

(1) the efficiency with which the carriers concerned are operated;
(2) the amount of revenue required by the carriers to pay their operating expenses and taxes and receive a fair return on the property held and used for transportaton;
(3) the importance of the transportation to the public;
(4) whether a particular participating carrier is an originating, intermediate, or delivering line; and
(5) other circumstances that ordinarily, without regard to the mileage traveled, entitle one carrier to a different proportion of a rate than another carrier.

It will be noted that 10705(b) applies to all carriers subject to the Commission's jurisdiction under Chapter 105 namely, railroads, motor carriers, water carriers and freight forwarders.

In *B. & M.R.R. v. U.S.*, 208 F. Supp. 661, the court said:

> "The prescribing of divisions is a legislative function, and the Commission's exercise of that power is conditioned upon its providing that the divisions then in force do not, or in the future will not, comply with the specified standards."

The revenue received on the majority of the traffic handled by the railroads is divided on percentages. They have been built up from year to year and for the most part are sufficiently flexible to divide all rates not only at the time of their establishment but also those subsequently revised. Some of the bases for the division of joint through rates are:

Mileage Prorates—When this basis is used, the number of miles each line handles a shipment governs the percentage of revenue received. Minima, the measure of which varies with different conditions, are generally observed and lines performing short hauls or those operating under unusual conditions are usually protected by inflated mileages or a minimum percentage of the revenues.

In *Official-Southern Divisions,* 287 ICC 497, the Commission had the following to say regarding mileage prorates:

> "The thought dominates the law, as it is now framed, that a paramount consideration in determining the equitable share of the joint revenue which any carrier shall receive must be the relative amount and cost of the service which it renders, and that mileage is not a gage of relative cost where terminal service is performed or the transportation is carried on under other unfavorable conditions."

Rate Prorates—The Interstate Commerce Commission has prescribed class rates in a large part of the United States and many of the percentages used are based upon the use of the first class rate of each line as factors. It was concluded that if local rates could be considered as the proper level to determine revenues on local movements, it was logical to use such rates as factors in the division of joint through rates. As an example, let us assume that the joint through rate from A to C is 90 cents via carrier X to junction B and carrier Y beyond. The local first class rate from A to B via X is $1.00 and the local first class rate from B to C via carrier Y is 50 cents. The sum of the two rates is $1.50, thus making the percentage of carrier X, 66 2/3 percent and that of carrier Y, 33 1/3 percent. Therefore, on traffic moving under the joint through rate of 90 cents, carrier X would receive 60 cents and carrier Y would receive 30 cents.

Specific Divisions—There are many special divisions in effect on basic commodities such as forest products, cotton, grain, coal, coke, ore, sand, etc. The rate structures on these commodities have considerable bearing on the manner in which the rates are divided. It is generally conceded that the services of the roads originating this character of traffic are greater than on other traffic and they are entitled to better earnings than the customary bases would produce.

Cost of service under economical and efficient management is therefore a vital factor in determining the fair and equitable share of joint revenue which each carrier should receive.

The Commission may, in the public interest, take into consideration the needs of a weaker road and accord it a division larger than justice merely between the parties would suggest, in order to maintain it in effective operation as part of an adequate transportation system, provided the share left to its connections is adequate to avoid a confiscatory result. *(D.S.S. & A. v. Mackinac Transp.,* 306 ICC 553.)

When divisions have been agreed upon between carriers, they are published as division sheets in as comprehensive yet simplified a manner as possible. Agents of the rate associations publish the divisions on interterritorial traffic and such issues have joint application and are compiled and printed by the associations whose carriers originate the traffic or have the larger interests. The individual carriers issue division sheets on intraterritorial traffic for local and joint application.

In the motor carrier industry there is no real uniformity in the method of dividing the revenue under through joint rates. In many instances, a rate prorate is used based on the rail Docket 28300, first class rates, rather than on the motor carrier class rates. In others, a motor carrier division factor is used, as well as Beranek's factor.

International Intermodal Through Routes and Joint Rates

For many years shipments moving from or to inland cities in the United States to or from inland cities in foreign countries have been charged a combination of rates made up of the inland rates to and/or from the ports plus the ocean freight rates charged by the steamship companies and any wharfage and handling charges applicable at the ports. The development and growth of containerization brought many changes. The so-called "break-bulk" operation (unloading of the freight from cars and trucks and loading into the ship) was eliminated and as a result of arrangements among the various carriers through service and through routes were established. Despite the existence of through routes combination rates continued to be charged with the inland movement, within the United States, subject to the jurisdiction of the Interstate Commerce Commission and the ocean transportation regulated by the Federal Maritime Commission.

Prior to April 1969, the ICC had taken the position that it had no authority to accept for filing tariffs establishing joint international rates between common carriers subject to its jurisdiction and ocean carriers subject to the jurisdiction of the Federal Maritime Commission. In response to congressional inquiries, and after an extensive review of the legislative history of the Interstate Commerce Act, the Commission on April 1, 1969 announced that it was now of the opinion that it did have authority to accept joint rail-ocean rates and that this jurisdiction also encompassed the filing of joint international rates between common carriers by motor vehicle and common carriers by water, on the one hand, and on the other, ocean carriers. Subsequently thereto, on July 15, 1969 the Commission by order in Special Permission No. 70-275 changed the existing regulations so as to permit common carriers by water (ocean carriers) to enter into arrangements with common carriers subject to the Commission's jurisdiction for the purpose of establishing joint rates and through routes. Following a protest by the Federal Maritime Commission the effectiveness of the ICC order was stayed pending further order of the ICC. On July 31, 1969 the Commission instituted a rule-making proceeding (Ex Part 261) with a view to amending its existing tariff rules pertaining to export and import traffic moving via common carriers subject to its jurisdiction and ocean-going common carriers operating in foreign commerce.

On September 4, 1970 the Commission announced its findings, *In the Matter of Tariffs Containing Joint Rates and Through Routes for the*

Transportation of Property between Points in the United States and Points in Foreign Countries (337 ICC 625), and adopted comprehensive regulations governing the filing of joint international rates including the requirement that tariffs containing such rates shall be published, filed and posted in conformity with the provisions of the Act and the rules of the Tariff Circular, shall include the names of all participating carriers, a description of the service to be performed by each participating carrier, and a statement of the divisions of the joint rate to be received by the participating carrier subject to the Commission's jurisdiction.

Since the original report in 1970 four supplemental reports have been issued by the Commission, June 28, 1972 (341 ICC 246), August 12, 1974 (346 ICC 688), June 29, 1975 (350 ICC 361), and January 30, 1976 (351 ICC 490). In view of the importance of this subject today the Commission's findings and order in the 1976 supplemental report are reproduced in their entirety.

"FINDINGS

"We find that §§1300.0 and 1300.67 of Subchapter D of Chapter X of Title 49 of the Code of Federal Regulations (49 CFR §§1300.0 and 1300.67), otherwise known as the preamble and rule 67 of Tariff Circular No. 20, should be amended as shown in the order attached hereto.

"We further find that §1305.0 of Subchapter D of Chapter X of Title 49 of the Code of Federal Regulations (49 CFR §1305.0), otherwise known as the preamble to Rule 33 of Tariff Circular No. 20, should be amended as shown in the order attached hereto.

"We further find that §1307.22 of Subchapter D of Chapter X of Title 49 of the Code of Federal Regulations (49 CFR §1307.22), otherwise known as the preamble of Tariff Circular MF No. 3, should be amended as shown in the order attached hereto.

"We further find that Subpart B of Part 1307 of Subchapter D of Chapter X of Title 49 of the Code of Federal Regulations, otherwise known as Tariff Circular MF No. 3, should be amended by the addition of a new §1307.49, as shown in the order attached hereto.

"We further find, that §1308.0 of Subchapter D of Chapter X of Title 49 of the Code of Federal Regulations (49 CFR §1308.0), otherwise known as the preamble of Tariff Circular No. 22, should be amended as shown in the order attached hereto.

"We further find, that the promulgation of the above-named new and revised rules necessitates the recision of Special Permission Nos. 70-275 and 73-123, and amendments thereto.

"We further find, that this decision is not a major Federal action significantly affecting the quality of the human environment within the meaning of the National Environmental Policy Act of 1969."

"At a General Session of the Interstate Commerce Commission, held at its office in Washington, D.C., on the 30th day of January 1976.

EX PARTE NO. 261
SPECIAL PERMISSION NO. 70-275
SPECIAL PERMISSION NO. 73-123

"In the Matter of Tariffs Containing Joint Rates and Through Routes for the Transportation of Property between Points in the United States and Points in Foreign Countries.

"ORDER

"Upon consideration of the record in the above-captioned proceeding, including the petitions for reconsideration of the report and order of the Commission on further reconsideration, served July 29, 1975, filed August 18, 1975, by the Chairman of the Federal Maritime Commission, and August 27, 1975, by IML Freight, Inc.; and

"It *appearing,* That, after review of the said petitions, the Commission has modified the regulations issued in the prior reports entered in this proceeding, the most recent report being printed at 350 ICC 361;

"Wherefore:

"It is ordered, That, pursuant to section 4 of the Administrative Procedure Act (5 U.S.C. §553), and sections 1(1)(a), 6(5), 6(6), 6(12), 202(a), 203(a)(11), 216(c), 217(a), 220(a), 302, 303, 305(b), 306, and 313 of the Interstate Commerce Act, in 49 CFR Chapter X,

"(1) Part 1300 be, and it is hereby, amended and revised as follows:

(a) Paragraph (a)(1) of §1300.0 (preamble to Tariff Circular No. 20) is amended to read as set forth in appendix A of the said report;

(b) §1300.67 (Rule 67 of Tariff Circular No. 20) is revised to read as set forth in appendix C of the said report;

"(The provisions of Part 1300 issued under sec. 12, 24 Stat. 383, as amended, 49 Stat. 546, as amended; 49 U.S.C. 12, 304, and secs. 5, 6, 24 Stat. 380, as amended, 49 Stat. 560, as amended; 49 U.S.C. 5, 6, 317.)

"(2) Part 1305 be, and it is hereby, amended as follows:

(a) §1305.0 (preamble to Rule 33 of Tariff Circular No. 20) is amended to read as set forth in appendix B of the said report;

"(The provisions of Subpart A of Part 1305 issued under sec. 6, 24 Stat. 380, sec. 12, 24 Stat. 383; 49 U.S.C. 6, 12.)

"(3) Part 1307 be, and it is hereby, amended and revised as follows:

(a) §1307.22 (preamble to Tariff Circular MF No. 3) is amended to read as set forth in appendix D of the said report:

(b) §1307.49 is added as a new section in the said part as set forth in appendix E of the said report;

"(The provisions of Subpart B of Part 1307 issued under secs. 204, 217, 49 Stat. 546, as amended, 560, as amended, sec. 210a, as amended, 52 Stat. 1238, as amended; 49 U.S.C. 304, 317, 310a.)

"(4) Part 1308 be, and it is hereby, amended and revised as follows: §1308.0 (preamble to Tariff Circular No. 22) is amended by changing paragraph (b) thereof to read as set forth in appendix F of the said report; and paragraph (c), is added as a new paragraph to the said section as set forth in appendix F of the said report:

"(The provisions of Part 1308 issued under secs. 304, 306, 54 Stat. 933, 935; 49 U.S.C. 904, 906.)

"It is further ordered, That the petitions be, and they are hereby, denied in all other respects for the reason that sufficient grounds have not been presented to warrant granting the action sought.

"It is further ordered, That Special Permission No. 70-275 be, and it is hereby, rescinded.

"It is further ordered, That Special Permission No. 73-123, and amendments thereto, be, and they are hereby, rescinded.

"It is further ordered, That all prior orders in Ex Parte No. 261 be, and they are hereby, vacated and set aside.

It is further ordered, That this order shall become effective 35 days from the date of service, and shall remain in effect until modified or revoked in whole or in part by further order of this Commission.

"And it is further ordered, That notice of this order shall be given to the general public by mailing a copy of this order to each party of record in this proceeding, by depositing a copy in the Office of the Secretary, Interstate Commerce Commission, Washington, D.C. 20423, for public inspection, and by delivering a copy to the Director, Office of the Federal Register.

"By the Commission."

"APPENDIX A

"§1300.0 GENERAL PROVISIONS; DEFINITIONS

"(a) *General application; conformation to rules; reissue.*

"(1) This part contains regulations issued by the Interstate Commerce Commission, under authority of section 6 of the Interstate Commerce Act, as amended, to govern the construction and filing of freight rate tariffs and classifications of railroads, water carriers and pipeline companies filing under section 6 of the act, and, under authority of sections 217 and 306 of the act, to govern the construction and filing of freight rate tariffs and classifications naming or governing joint rates and routes over motor and water carriers jointly with such carriers subject to part I of the act. The regulations in this part shall also govern the construction and filing of tariffs naming through routes and joint rates over the lines of common carriers by railroad, water, or pipeline, or by railroad jointly with common carriers by motor vehicle, subject to the Interstate Commerce Act, on the one hand, and vessel-operating common carriers by water engaged in the foreign commerce of the United States, as defined in the Shipping Act, 1916, on the other hand, for the transportation of property between any place in the United States and any place in a foreign country. See §1300.67.

"APPENDIX B

"§1305.0 APPLICATION—POSTING OF TARIFFS DEFINED

"(a) The regulations in this subpart shall also govern the posting (by carriers subject to the jurisdiction of the Interstate Commerce Commission) of any tariff containing a through route and joint rate over the lines of a common carrier by railroad, pipeline, or water, or by railroad jointly with a common carrier by motor vehicle, subject to the Interstate Commerce Act, on the one hand, and a vessel-operating common carrier by water engaged in the foreign commerce of the United States, as defined in the Shipping Act, 1916, on the other hand, and all other tariffs governing the application of the rate tariff, for the transportation of property between any place in the United States and any place in a foreign country. The carrier subject to the jurisdiction of this Commission receiving shipments at a port for delivery to points in the United States under joint through rate and route arrangements shall post at its station at such port the tariffs naming such rates and its governing tariffs. See §1300.67.

"(b) The term 'post; as used in this part means the maintenance of a file of tariffs in the custody of an agent of the carrier in a complete, accessible, and usable form, and keeping such file of tariffs available to the public upon request during ordinary business hours. The term 'tariff' as used in this part includes tariff supplements or amendments.

"APPENDIX C

"§1300.67 EXPORT AND IMPORT TRAFFIC—OCEAN CARRIERS

"(a) *Ocean carriers not subject to Act.* Common carriers by water, or conferences of such carriers, engaged in the foreign commerce of the United States, as defined in the Shipping Act, 1916, that operate between ports of the United States and foreign countries are not subject to the terms of the Interstate Commerce Act or to the jurisdiction of the Interstate Commerce Commission.

"(b) *Through routes and joint rates.*

"(1) A common carrier by railroad, pipeline, or water, or a common carrier by railroad jointly with a common carrier by motor vehicle, subject to the Interstate Commerce Act (hereinafter referred to in this section as the domestic carrier), may establish a through route and joint rate with a vessel-operating common carrier by water engaged in the foreign commerce of the United States (hereinafter referred to in this section as the ocean carrier), as defined in the Shipping Act, 1916, for the transportation of property between any place in the United States and any place in a foreign country. Every tariff naming such a through route and joint rate shall be filed with this Commission. The tariff may be filed in the name of the ocean carrier, a conference of ocean carriers, the domestic carrier or the duly appointed tariff publishing agent of such carriers.

"(2) The tariff shall be constructed, filed, and posted in conformity with the Interstate Commerce Act, and, except as otherwise specifically authorized, with the regulations in Parts 1300 and 1305 (regulations in both parts included in Tariff Circular No. 20) of this chapter. The tariff shall be printed in the English language, include the names of all participating carriers, a description of the services to be performed by each participating carrier, a statement of the joint rate, and a clear and definite statement of the division, rate, or charge to be received by the domestic carrier for its share of the revenue covering a through shipment or aggregate of shipments under the tariff. The division, rate, or charge accruing to the domestic carrier must be shown in terms of lawful money of the United States. If shipments and/or loaded containers are to be permitted to be aggregated which are rated under more than one tariff published by the carrier or for its account, each tariff so affected must contain a specific rule, providing for the aggregation in connection with the statement of the domestic carrier's divisions and identifying by ICC designation each of the other tariffs. A tariff filed in the name of a conference need not show 'Agent' after the name of the conference unless the conference publishes as an agent. If a tariff provides less-than-carload, less-than-containerload, or less-than-trailerload service, such service must be defined. If the tariff provides containerload rates, such rates must be made subject to a specified minimum weight or minimum measure per container, or a specified minimum charge per shipment per container, and a maximum weight per container. Where the freight is to be packed (loaded) or unpacked (unloaded) into or from the containers by the domestic carrier, the tariff must clearly state that the joint rate includes this service or must provide a separate charge to apply when said service is provided.

"(3) Rates or charges may be stated to apply in a unit other than a United States unit provided the unit is defined in the tariff where used. The International System of Units (SI) (the metric system) may be used and need not be defined. A rate or charge applying on a unit of measurement other than weight may be published, but if the tariff also includes a rate or charge applying on a unit of weight on the same traffic, the charges on the weight basis must alternate with the charges on the measurement basis other than weight. In every case the tariff shall provide a definite method for determining the measurement of the shipment and the applicable charges. 'Cargo, N.O.S.' may be provided as a commodity description provided the term is clearly defined in the tariff where used. Tariffs governing the application of the rate tariff need not show a carrier as a participant when none of the provisions therein apply for such carrier's account.

"(4) Allowances, cargo administrative charges, or reductions shall not be provided for payment to shippers or other parties for services performed by or facilities furnished by other than the carriers parties to the through transportation unless (i) such carriers by tariff publication hold themselves out to perform such services and furnish such facilities, (ii) such carriers are able to perform such services and furnish such facilities upon reasonable demand, and (iii) the performance of such services and furnishing of such facilities are included in the through joint rate or charge. This subparagraph does not apply where such provisions do not affect the division, rate, or charge accruing to the domestic carrier or the services performed by such carrier.

"(5) A domestic carrier desiring to become a participant in a tariff filed in the name of a conference of ocean carriers, which conference does not publish as an agent, must give to its connecting ocean carrier participating in such conference tariffs a concurrence in tariffs issued and filed by the ocean carrier or the conference, or both. A limited concurrence may provide for only those limitations authorized in §1300.19 of this chapter. The concurrence forms prescribed by §1300.19 shall be modified to show that the authority extends to amendments to the tariff(s) and extends to tariffs filed in the name of the conference, and to show the types of tariffs (such as tariffs containing joint rail-ocean rates, joint rail-motor-ocean rates, et cetera) in which the domestic carrier desires to participate. Powers of attorney must not be executed unless the conference publishes as an agent.

"(6) The following changes may be published to become effective upon a specified date not prior to the date filed with the Commission in

Washington, D.C., provided the division, rate, or charge accruing to the domestic carrier or a provision governing or affecting such division, rate, or charge does not change.

"1. A change in a published rate, charge, rule, regulation, or other provision which results in a reduction or in no change in charges. This includes a change in a rate or charge which results in lessening or canceling a proposed (published but not yet effective) increase.

"2. The establishment of a rate on a specific commodity not previously named in a tariff which results in a reduction or in no change in charges. The tariff must contain a cargo, N.O.S. rate or similar general cargo rate, which rate would otherwise be applicable to the specific commodity. The specific commodity rate must be equal to or lower than the cargo, N.O.S. or general cargo rate.

"Except as otherwise provided in this subparagraph, no new or initial rate, charge, rule, regulation, or other provision and no new point of origin or destination may be published upon less than 30 days' notice. In no case may the establishment of or a change in a division, rate, or charge accruing to the domestic carrier or a provision governing or affecting such division, rate, or charge become effective upon less than 30 days' notice.

"(7) If a tariff includes charges for terminal services, canal tolls, or additional charges not under the control of the carrier or conference, which carrier merely acts as a collection agent for the charges, and the agency making such charges to the carrier increases the charges without notice or without adequate notice to the carrier or conference, such charges may be increased in the tariff by specific publication effective upon a specified date not prior to the date filed with the Commission, in Washington, D.C., whether included in the joint rate or separately stated. If the change occurs in the division, rate, or charge accruing to the domestic carrier, the amendment must contain a statement explaining the change.

"(8) Every change made under authority of §1300.67(b)(6) or (7) must be shown in an amendment (a supplement if the tariff is in bound form or a loose-leaf page if the tariff is in loose-leaf form) to the tariff. The rates, charges, rules, regulations, or other provisions authorized to be changed thereunder may be changed without their having been effective for 30 days prior to the effective date of the change.

"(9) The regulations in §1300.9(k) of this chapter—Suspension of Tariff Schedules—shall govern only when the operation of the division, rate, or charge accruing to the domestic carrier or any provision governing the division, rate, or charge or the service performed by such carrier is suspended by an order of this Commission.

"(10) The following reference marks may be used in the exact form shown for the purposes indicated and may not be used for any other purpose:

"(R) to denote reductions.

"(A) to denote increases.

"(C) to denote changes in wording which result in neither increases nor reductions in charges.

"An explanation of these reference marks must be provided in the tariff in which used.

"(c) *Port combination basis.* Domestic and ocean carriers may enter into joint rate arrangements, as authorized by paragraph (b) of this section, and domestic carriers may at the same time maintain in effect rates applicable only from and to the ports, usable in combination with ocean carriers' independently established rates. Publication of such rates by the domestic carrier shall be subject to the following:

"(1) The domestic carriers shall file their rates to the ports and from the ports, and such rates must be the same for all, regardless of which ocean carrier may be designated by the shipper, except as otherwise provided by section 28 of the Merchant Marine Act (41 Stat. 988, 46 U.S.C. 884).

"(2) When the domestic carriers publish rates which are indicated to apply only on export or import traffic, the tariffs containing such rates shall specify by inclusion or exclusion the countries to or from which traffic subject to such rates shall move, regardless of whether such countries are, or are not, adjacent to the United States. Tariffs shall also specify whether or not property destined to or coming from the Republic of Cuba, the Commonwealth of Puerto Rico, Guam, Hiawaii, or the Canal Zone is subject to such rates. In the absence of a statement in tariffs limiting the application of export or import rates, such rates will apply on traffic destined to or coming from them.

"(3) As a matter of convenience to the public, the domestic carriers may also publish as information in their tariffs the ocean carriers' rates or charges that will apply to or from a foreign country in connection with the domestic carriers' rates. When this is done, the ocean carriers' rates or charges are in no manner subject to the jurisdiction of this Commission, but the rates of the domestic carriers applying to or from the ports are subject to all provisions of the Interstate Commerce Act and to this Commission's regulations.

"(d) *Through export and import billing.* Export and import shipments may be forwarded under through billing. Through bills of lading must clearly separate the liability of the carriers included therein, where different, and must show (1) the tariff rates or charges of the domestic carriers to or from the port or (2) the joint rates or charges when such rates or charges are established and are named in tariffs on file with this Commission as provided in paragraph (b) of this section. The name of the domestic carrier shall appear in a prominent place on the face of the bill of lading when that carrier originates the shipment. Tariffs which provide for the use of a specified kind of bill of lading shall reproduce all of the terms and conditions thereof.

"Cross Reference: For regulations governing the posting of freight tariffs of common carriers by rail, water, and pipeline, including tariffs containing joint motor-rail or motor-rail-water rates, see Subpart A of Part 1305 of this chapter.

"APPENDIX D

"§1307.22 APPLICATION OF REGULATIONS

"(a) The regulations in Subpart B will also apply to tariffs containing joint rates of common carriers of property by motor vehicle and common carriers by water subject to Part III of the Interstate Commerce Act, other than railroad-owned or railroad-controlled water carriers.

"(b) The regulations in Subpart B will also apply to tariffs containing through routes and joint rates over the lines of common carriers by motor vehicle or by motor vehicle jointly with common carriers by water whether or not the water carriers are railroad-owned or railroad-controlled, subject to the Interstate Commerce Act, on the one hand, and vessel-operating common carriers by water engaged in the foreign commerce of the United States, as defined in the Shipping Act, 1916, on the other hand, for the transportation of property between any place in the United States and any place in a foreign country. See §1307.49.

"(c) The regulations in Subpart B will not apply (i) to tariffs containing joint rates between motor carriers, on the one hand, and, on the other hand, common carriers by rail or by water when such water carriers are railroad-owned or railroad-controlled and operate under the provisions of section 5(16) of the Interstate Commerce Act; or (ii) to tariffs containing joint motor-rail-water rates whether or not the water carrier is railroad-owned or railroad-controlled. See paragraph (b) of this section for exception as to ocean carriers.

"APPENDIX E

"§1307.49 EXPORT AND IMPORT TRAFFIX—OCEAN CARRIERS

"(a) *Ocean carriers not subject to Act.* Common carriers by water, or conferences of such carriers, engaged in the foreign commerce of the United States, as defined in the Shipping Act, 1916, that operate between ports of the United States and foreign countries are not subject to the terms of the Interstate Commerce Act or to the jurisdiction of the Interstate Commerce Commission.

"(b) *Through routes and joint rates.*

"(1) A common carrier by motor vehicle or by motor vehicle jointly with a common carrier by water, subject to the Interstate Commerce Act (hereinafter referred to in this section as the domestic carrier), may establish a through route and joint rate with a vessel-operating common carrier by water engaged in the foreign commerce of the United States (hereinafter referred to in this section as the ocean carrier), as defined in the Shipping Act, 1916, for the transportation of property between any place in the United States and any place in a foreign country. Every tariff naming such a through route and joint rate shall be filed with this Commission. The tariff may be filed in the name of the ocean carrier, a conference of ocean carriers, the domestic carrier or the duly appointed tariff publishing agent of such carriers.

"(2) The tariff shall be constructed, filed, and posted in conformity with the Interstate Commerce Act, and, except as otherwise specifically authorized, with the regulations in Subpart B of Part 1307 (Tariff Circular MF No. 3) of this chapter. The tariff shall be printed in the English language, include the names of all participating carriers, a description of the services to be performed by each participating carrier, a statement of the joint rate, and a clear and definite statement of the division, rate, or charge to be received by the domestic carrier for its share of the revenue covering a through shipment or aggregate of shipments under the tariff. The division, rate, or charge accruing to the domestic carrier must be shown in terms of lawful money of the United States. If shipments and/or loaded containers are to be permitted to be aggregated which are rated under more than one tariff published by the carrier or for its account, each tariff so affected must contain a specific rule, providing for the aggregation in connection with the statement of the domestic carrier's divisions and identifying by ICC designation each of the other tariffs. A tariff filed in the name of a conference need not show 'Agent' after the name of the conference unless the conference publishes as an agent. If a tariff provides less-than-truckload, less-than-containerload, or less-than-trailerload service, such service must be defined. If the tariff provides containerload rates, such rates must be made subject to a specified minimum weight or minimum measure per container, or a specified minimum charge per shipment per container, and a maximum weight per container. Where the freight is to be packed (loaded) or unpacked (unloaded) into or from the containers by the domestic carrier, the tariff must clearly state that the joint rate includes this service or must provide a separate charge to apply when said service is provided.

"(3) Rates or charges may be stated to apply in a unit other than a United States unit provided the unit is defined in the tariff where used. The International System of Units (SI) (the metric system) may be used and need not be defined. A rate or charge applying on a unit of measurement other than weight may be published, but if the tariff also includes a rate or charge applying on a unit of weight on the same traffic, the charges on the weight basis must alternate with the charges on the measurement basis other than weight. In every case the tariff shall provide a definite method for determining the measurement of the shipment and the applicable charges. 'Cargo, N.O.S.' may be provided as a commodity description provided the term is clearly defined in the tariff where used. Tariffs governing the application of the rate tariff need not show a carrier as a participant when none of the provisions therein apply for such carrier's account.

"(4) Allowances, cargo administrative charges, or reductions shall not be provided for payment to shippers or other parties for services performed by or facilities furnished by other than the carriers parties to the through transportation unless (i) such carriers by tariff publication hold themselves out to perform such services and furnish such facilities, (ii) such carriers are able to perform such services and furnish such facilities upon reasonable demand, and (iii) the performance of such services and furnishing of such facilities are included in the through joint rate or charge. This subparagraph does not apply where such provisions do not affect the division, rate, or charge accruing to the domestic carrier or the services performed by such carrier.

"(5) A domestic carrier desiring to become a participant in a tariff filed in the name of a conference of ocean carriers, which conference does not publish as an agent, must give to its connecting ocean carrier participating in such conference tariffs a concurrence in tariffs issued and filed by the ocean carrier or the conference, or both. A limited concurrence may provide for only those limitations authorized in §1307.47(c) of this chapter. The concurrence forms prescribed by §1307.47(c) shall be modified to show that the authority extends to amendments to the tariff(s) and extends to tariffs filed in the name of the conference, and to show the types of tariffs (such as tariffs containing joint motor-ocean rates, et cetera) in which the domestic carrier desires to participate. Powers of attorney must not be executed unless the conference publishes as an agent.

"(6) The following changes may be published to become effective upon a specified date not prior to the date filed with the Commission in Washington, D.C., provided the division, rate, or charge accruing to the domestic carrier or a provision governing or affecting such division, rate, or charge does not change.

"1. A change in a published rate, charge, rule, regulation, or other provision which results in reduction or in no change in charges. This includes a change in a rate or charge which results in lessening or canceling a proposed (published but not yet effective) increase.

"2. The establishment of a rate on a specific commodity not previously named in a tariff which results in a reduction or in no change in charges. The tariff must contain a cargo, N.O.S. rate or similar general cargo rate, which rate would otherwise be applicable to the specific commodity. The specific commodity rate must be equal to or lower than the cargo, N.O.S. or general cargo rate.

"Except as otherwise provided in this subparagraph, no new or initial rate, charge, rule, regulation, or other provision and no new point of origin or destination may be published upon less than 30 days' notice. In no case may the establishment of or a change in a division, rate, or charge accruing to the domestic carrier or a provision governing or affecting such division, rate, or charge become effective upon less than 30 days' notice.

"(7) If a tariff includes charges for terminal services, canal tolls, or additional charges not under the control of the carrier or conference, which carrier merely acts as a collection agent for the charges, and the agency making such charges to the carrier increases the charges without notice or without adequate notice to the carrier or conference, such charges may be increased in the tariff by specific publication effective upon a specified date not prior to the date filed with the Commission, in Washington, D.C., whether included in the joint rate or separately stated. If the change occurs in the division, rate, or charge accruing to the domestic carrier, the amendment must contain a statement explaining the charge.

"(8) Every change made under authority of §1307.49(b)(6) or (7) must be shown in an amendment (a supplement if the tariff is in bound form or a loose-leaf page if the tariff is in loose-leaf form) to the tariff. The rates, charges, rules, regulations, or other provisions authorized to be changed thereunder may be changed without their having been effective for 30 days prior to the effective date of the change.

"(9) The regulations in §1307.34 of this chapter—Suspension of Tariff Schedules—shall govern only when the operation of the division, rate, or charge accruing to the domestic carrier or any provision governing the division, rate, or charge or the service performed by such carrier is suspended by an order of this Commission.

"(c) *Port combination basis.* Domestic and ocean carriers may enter into joint rate arrangements, as authorized by paragraph (b) of this section, and domestic carriers may at the same time maintain in effect rates applicable only from and to the ports, usable in combination with ocean carriers' independently established rates. Publication of such rates by the domestic carrier shall be subject to the following:

"(1) The domestic carriers shall file their rates to the ports and from the ports, and such rates must be the same for all, regardless of which ocean carrier may be designated by the shipper except as otherwise provided by section 28 of the Merchant Marine Act (41 Stat. 988, 46 U.S.C. 884).

"(2) When the domestic carriers publish rates which are indicated to apply only on export or import traffic, the tariffs containing such rates shall specify by inclusion or exclusion the countries to or from which traffic subject to such rates shall move, regardless of whether such countries are, or are not, adjacent to the United States. Tariffs shall also specify whether or not property destined to or coming from the Republic of Cuba, the Commonwealth of Puerto Rico, Guam, Hawaii, or the Canal Zone is subject to such rates. In the absence of a statement in tariffs limiting the application of export or import rates, such rates will apply on traffic destined to or coming from them.

"(3) As a matter of convenience to the public, the domestic carriers may also publish as information in their tariffs the ocean carriers' rates or charges that will apply to or from a foreign country in connection with the domestic carriers' rates. When this is done, the ocean carriers' rates or charges are in no manner subject to the jurisdiction of this Commission, but the rates of the domestic carriers applying to or from the ports are subject to all provisions of the Interstate Commerce Act and to this Commission's regulations.

"(d) *Through export and import billing.* Export and import shipments may be forwarded under through billing. Through bills of lading must clearly separate the liability of the carriers included therein, where different, and must show (1) the tariff rates or charges of the domestic carriers to or from the port, or (2) the joint rates or charges when such rates or charges are established and are named in tariffs on filed with this Commission as provided in paragraph (b) of this section. The name of the domestic carrier shall appear in a prominent place on the face of the bill of lading when that carrier originates the shipment. Tariffs which provide for the use of a specified kind of bill of lading shall reproduce all of the terms and conditions thereof.

"APPENDIX F

"§1308.0

"(b) *Conformation to regulations, et cetera.* All tariffs applicable on traffic subject to the regulations in this subpart must conform to all of their provisions, except as otherwise authorized by the Commission. They shall state and arrange the rates, charges, rules and regulations clearly and explicitly in such manner that there will be no doubt of their proper application. Carriers or their agents may not publish rates or provisions which duplicate or conflict with rates or provisions published by or for account of such carriers.

"(c) *Nonapplication.* Tariffs governed by the regulations in this subpart must not contain:

"Joint water-rail rates;

"Joint water-motor rates;

"Joint water-rail-motor rates;

"Passenger fares;

"Contract carrier rates or charges.

"Furthermore, tariffs published under this subpart must not contain joint rates and through routes over the lines of common carriers by water subject to the Interstate Commerce Act, on the one hand, and vessel-operating common carriers by water engaged in the foreign commerce of the United States, as defined in the Shipping Act, 1916, on the other hand, for the transportation of property between any place in the United States and any place in a foreign country. For applicable regulations, see §§1300.67 and 1307.49."

CHAPTER 5

ARBITRARIES AND DIFFERENTIALS

General

The words "arbitrary" and "differential" are frequently used interchangeably and there are many instances in which their application is exactly the same. It has been said that arbitraries are "added to" and differentials are "deducted from" a base rate and although true in many cases, it is not uniformly so. For example:

(a) An arbitrary is a fixed amount added to or deducted from a base rate to make a rate from or to another point.

(b) A differential is the difference established between rates from related points of origin, or to related points of destination, or via different routes between the same points.

(c) A differential rate is a rate established via a particular route or mode of transportation by deducting a fixed amount from or adding a fixed amount to the rate via another (standard) route or mode of transportation between the same points.

(d) A differential route is one via which differential rates are maintained. It is usually one that is more circuitous or less desirable, for one reason or another, and via which lower rates must be maintained in order to attract business.

The use of arbitraries and differentials in rate making is of long standing. In the early days, prior to the creation of the Commission a great many joint through rates were made by the use of arbitraries or differentials. The Commission, however, has never favored this method of making rates but has recognized the necessity of using arbitraries or differentials in many rate situations and has approved their use in many adjustments.

Arbitraries and differentials are established only when the necessities of commerce require it. In each section of the country it has been found necessary from time to time to employ arbitraries or differentials and to establish rate structures predicated thereon so as to maintain a proper relationship between the respective territories, rate groups, sections, rate scales, commodities and routes.

Rule 15 of Tariff Circular 20 describes the information which must be shown in rate basis books, and with regard to arbitraries and differentials states in part as follows:

> "The rate basis, group location, or territorial application, or arbitraries or differentials to be added to or deducted from the base or group rates, must be shown immediately in connection with the name of each station, except that reference may there be made to an item showing such information."

Tariff Circular MF No. 5 contains a rule similar to Rule 4(i)(4) of Circular No. 20.

Under certain conditions, motor carriers have found it advisable to publish arbitraries. In areas where traffic density is very light through rates have been published to and from key points and arbitraries are then added to make rates to and from other points in the area. Such arbitraries are justified on the ground that the size and number of shipments moving to and from these points are small and adequate revenues cannot be obtained through the normal scale of rates.

Congestion at piers and wharves also causes motor carriers to publish arbitraries on shipments to and from such locations. The unusually long periods of waiting time which sometimes occur in picking up and delivering shipments at piers and wharves can make this a high cost operation and the assessment of arbitraries is designed to cover part or all of the costs resulting from these conditions.

Some indication of the attitude of the Commission with regard to arbitraries may be obtained from the following comment in *Less than Truckload Arbitraries from and to Minnesota points,* 309 I.C.C. 527 (1960):

> "Proposed 16 cent arbitrary, to be added to existing rates and charges on interstate LTL traffic originated or delivered by respondents at 58 Minnesota points, is not justified when it has no relation to distance or specific handling costs, would fail to maintain existing percentage relations to first class and would be unjustly discriminatory."

Some of the principal reasons why differentials and arbitrarties are established and maintained are:

1—To maintain agreed relationships between producing and/or distributing centers.

2—To maintain equitable relationships between rate structures.

3—To maintain equitable relationships between communities subject to a geographical disadvantage as compared to competitive communities more favorably situated.

4—To enable lines with a low traffic density to compete with lines having a heavy traffic density.

5—To make it possible for circuitous lines, water carriers, rail and water routes to obtain a fair share of the traffic between points competitive with the more direct lines or more desirable routes.

6—To enable weak or short lines to obtain a better share of traffic and earnings.

7—To adequately compensate a carrier or carriers for high terminal costs in congested, heavy traffic areas such as New York City.

Arbitraries

In every major railroad class rate adjustment since 1925 the Commission has prescribed arbitraries of one kind or another. In most cases they have been built into the rate structure and are not shown separately in the class rate tariffs. In such cases the user of the tariff has no way of knowing that the class rate scale is constructed by combining a base scale with certain prescribed arbitraries except by referring to the decision of the Commission. Where the arbitraries, either class or commodity, are published separately in the rate tariff such publication would be subject to the rules of Tariff Circular 20. Rule 4(i)(4) of Supplement 8 of the Circular states:

> "A tariff may provide rates from or to designated points by the addition or deduction of arbitraries or differentials, from or to rates shown therein from or to base points, but provision for the addition or deduction of arbitraries or differentials must be shown either in connection with the base rate or in a separate item which must specifically name the base point. (See rule 15.)"

While a rate made by the use of an arbitrary added to the base rate might, on the face of it, appear to be a combination rate, the Commission has ruled otherwise. In *Brainerd Fruit Co., v. C.G.W. Ry.*, 163 I.C.C. 585 the Commission has held that rates published as arbitraries over base points are joint through rates and not combination rates. In this connection it should be noted that in attacking the reasonableness of such joint through rates it is usually necessary to attack the through rate as a whole rather than one or another of the separate factors making up the the through rate. On this subject the Commission in *Acme Fast Freight v. Western Freight Association*, 299 I.C.C. 315, stated:

"While it is ordinarily true that the test of reasonableness of through rates comprised of arbitraries and terminal-to-terminal rates as in the case of the defendant's rate structure, should be applied to the rates or charges as a whole, and not to the separate factors, this principle may not be employed to cloak a scheme of ratemaking under which the defendant, in order to attract traffic from off-line points, affords to shippers non-compensatory arbitrary rates. To provide shippers with a service of value without being adequately compensated therefor, and as an inducement to secure their traffic, is an indirect form of illegal rebating."

Differentials

Agreed differentials established by the railroads actually antedate the Act to Regulate Commerce. One of the earliest was the establishment of a differential relationship between the New York, Philadelphia and Baltimore rates on grain from Chicago. The port differentials established in the late 1870's were continued for many years. (For a discussion of the port differentials see Chapter 2.)

In addition to those established to the Atlantic ports differential rates were also established via lake and rail routes between the east and the west. These rates were made differentially under the standard all rail routes. Likewise on traffic between the Atlantic seaboard and western points routed via the more circuitous rail routes (through Canada) rates were established on the basis of differentials under those applicable via standard routes. Rates differentially under the standard all rail rates were also established via rail-barge routes from the Midwest to Gulf ports.

Although the original port differentials were the result of an agreement between the rail carriers prior to the creation of the I.C.C. many differential relationships have not only been approved by the Interstate Commerce Commission but also prescribed by them. In *Commodities over Tidewater Express Lines,* 2 M.C.C. 356 the Commission approved differentials between motor-water rates and all motor rates. Also in *Increased Common Carrier Truck Rates in East,* 42 M.C.C. 633 the Commission stipulated that the increase should be applied to the New York City Zone 1 rates and the existing differentials to the other New York City zones preserved. On the other hand in *Motor Carrier Rates in New England,* 8 M.C.C. 287 they found that there was no justification for differentials in favor of small carriers or between regular and irregular route carriers.

Traditionally water carrier rates have been maintained differentially under the all rail rates between the same points. However, under the Supreme Court's interpretation of the following provisions of the Interstate

Commerce Act only extraordinary circumstances can justify the establishment and maintenance of such differentials.

Section 10705(a)(1) of the Act states in part:

> "When one of the carriers on a through route is a water carrier, the Commission shall prescribe a differential between an all-rail rate and a joint rate related to the water carrier if the differential is justified."

The wording of Section 10705(a)(1) would seem to indicate that Congress is not merely saying that the Commission may authorize such differentials where common carriers by water are involved but that this must be done. No mention is made, however, of all-water routes versus all-rail routes. On the other hand Section 10704(a)(2) of the statute provides, "However, a rate, classification, rule or practice of a rail carrier may be maintained at a particular level to protect the traffic of another carrier or mode of transportation only if the Commission finds that the rate or classification, or rule or practice related to it reduces or would reduce the going concern value of the carrier charging the rate."

The National Transportation Policy reads as follows:

> "(a) To ensure the development, coordination, and preservation of a transportation system that meets the transportation needs of the United States, including the United States Postal Service and national defense, it is the policy of the United States Government to provide for the impartial regulation of the modes of transportation subject to this subtitle, and in regulating those modes—
>
> (1) to recognize and preserve the inherent advantage of each mode of transportation;
>
> (2) to promote safe, adequate, economical, and efficient transportation;
>
> (3) to encourage sound economic conditions in transportation, including sound economic conditions among carriers;
>
> (4) to encourage the establishment and maintenance of reasonable rates for transportation without unreasonable discrimination or unfair or destructive competitive practices;
>
> (5) to cooperate with each State and the officials of each State on transportation matters; and
>
> (6) to encourage fair wages and working conditions in the transportation industry.
>
> (b) This subtitle shall be administered and enforced to carry out the policy of this section."

The circumstances leading to the Supreme Court's decision on the subject of differential rates for water carriers are as follows: in 1960 the Commission, *In Commodities—Pan Atlantic S.S. Corp., I & S M10415,*

309 I.C.C. 587, approved rates from points in the east to points in the southwest via the sea-land service of Pan Altantic S.S. lower than the all-rail routes between the same points. The rail lines countered by establishing TOFC rates on a parity with the sea-land rates. These reduced rates were suspended by the Commission. In their decision, 313 I.C.C. 23, they held that although compensatory, the rates should not be approved. They stated that TOFC rates should be no lower than 6% above the rates of Pan Atlantic Steamship and that box car rates should be somewhat less than 6%.

Following the Commission's decision several railroads brought an action in the U.S. District Court to enjoin the order of the Commission in *Commodities—Pan Atlantic Steamship Corp.*, 313 I.C.C. 23. The District Court in *N.Y.N.H.&H. R.R. v. U.S.*, 199 F. Supp. 635 (1961) ruled that the I.C.C. order requiring cancellation of the TOFC rates should be set aside and the I.C.C. enjoined from cancellation of TOFC rates which return at least the fully-distributed cost of carriage. Upon appeal the United States Supreme Court in *I.C.C. v. N.Y.N.H.&H. R.R.*, 372 U.S. 744 (1963) agreed with the District Court that the Commission's order must be set aside and ordered the judgment below vacated and the case remanded to the Commission for further proceedings consistent with their opinion. In their opinion, the court stated:

> "We agree with the District Court that 'at least on this record' the Commission's rejection of the TOFC rates here at issue and the requirement of a differential over the rates of the coastwise carriers were not consistent with the mandate of Section 15a(3). In light of the findings and conclusions underlying the Commission's decision, and more particularly its putting aside the question of 'inherent advantages,' its insistence that TOFC rates, in the words of the prohibition in Section 15a(3), 'be held up to a particular level to protect the traffic' of the coastwise carriers cannot be justified on the basis of the objectives of the National Transportation Policy."

Methods of Publishing Arbitraries and Differentials

To illustrate a method used in publishing arbitraries, portions of two pages taken from Middlewest Motor Freight Bureau Tariff 535-C are reproduced on pages 87 and 88. This tariff publishes class rates between points in Middlewest Territory and points in Southwestern Territory. Assuming a shipment of 2,500 pounds rated Class 70 from Somers, Iowa to Dallas, Texas, we find by referring to the Alphabetical List of Points in Middlewest Territory that Somers takes Group 150 and in addition refers to Item 160. This item has been reproduced and in it we find that various arbitraries are provided. On the shipment in ques-

Tariff MWB 535-C

ALPHABETICAL LIST OF POINTS IN MIDDLEWEST TERRITORY, BY STATES, SHOWING GROUP NUMBER APPLICABLE

POINT	GROUP NO.	POINT	GROUP NO.	POINT	GROUP NO.
POINTS IN IOWA					
Quarry	142	Saylorville.	148	Somers (See Item 160). .	150
Quick.	168	Scarville (See Item 160)	174	South Amana.	144A
Quimby..	160	Schaller (See Item 160).	152	South English.	144A
Radcliff (See Item 160).	150A	Schleswig (See Item 160)	152	Spencer.	158
Ralston (See Item 160) .	148B	Scranton (See Item 160).	148B	Sperry (See Item 160). .	26X
Randall.	282A	Seney (See Item 160) . .	160	Spillville (See Item 162).	164
Randolph	282A	Seargeant Bluff (Sergeant Bluff). . . .	284B	Spragueville (See Item 162).	146
Raymond (See Item 162) .	156A	Secton	158	Springbrook (See Item 162).	146
Read (See Item 162). . .	164	Seymour.	132	Springdale (See Item 162).	144A
Readlyn (See Item 162) .	156A	Shaffton	30X	Springville (See Item 162).	156
Redfield	148A	Shaffton Station	30X	Spirit Lake.	166
Red Oak.	128	Shambaugh (See Item 160)	128	Stacyville (See Item 160).	
Reeder Mills	148C	Sharon	132		
Reever	162C	Sharpsburg	128		
Reinbeck (See Item 160).	154	Sheffield (See Item 160)	162		
Rembrandt (See Item 160)	160	Shelby (See Item 160). .	138		
Remsen (See Item 160). .	160A				
nwick (See Item 160) .	158A				

tion it would be necessary to assess an arbitrary of 32 cents per 100 pounds which would be added to the charges based on the basic class scale between Somers and Dallas. If the shipment were destined to a point in Texas which also took an arbitrary this would also have to be added. The freight bill would then show two or three separate charges, (1) the arbitrary charge at Somers, (2) the charge based on the Class 70 rate and (3) the arbitrary charge, if any, applying at the destination.

Another method of publishing arbitraries, one employed by the rail carriers, is found in Transcontinental Freight Bureau Tariff No. 3001. This tariff names commodity rates from points in the east, south and midwest to points on the west coast and also to points in Arizona and New Mexico. To numerous points in Arizona and New Mexico it is necessary to add an arbitrary from some base point to the ultimate destination. The territorial directory which must be used in conjunction with TCFB Tariff No. 3001 to find the origin group and the destination rate basis indicates that rates to a point such as Globe, Arizona will be arrived at by adding an arbitrary from Bowie, Arizona. The Table of Contents in the rate tariff refers to two items for Arizona and New Mexico arbitraries. In one item an arbitrary scale number is provided applying from Bowie to Globe and the other item names an arbitrary in cents per 100 pounds which must be assessed in addition to the commodity rate to the basing point, Bowie.

Application of Rates

Item	Application
160	Arbitraries From Or To Iowa Points Referring Hereto

Rates or charges on shipments subject to LTL or AQ rates from or to points referring to this item will be made by adding the arbitrary rates or charges shown below: (See Notes 1, 2, 3 and 4).

Shipments Subject To	Arbitrary in Cents Per 100 Pounds
LTL or AQ rates 1 pound and over, but less than 1,000 pounds.	47
LTL or AQ rates 1,000 pounds and over, but less than 2,000 pounds.	41
LTL or AQ rates 2,000 pounds and over, but less than 3,000 pounds.	32
LTL or AQ rates 3,000 pounds and over, but less than 5,000 pounds.	26
LTL or AQ rates 5,000 pounds an- over, but less than 10,000 pounds.	19
LTL or AQ rates 10,000 pounds and over.	12

NOTE 1—Subject to a minimum arbitrary charge of 190 cents.

NOTE 2—The provisions of this item will not apply from or to Eldora, IA to the following extent:

(1) On shipments destined to Eldora, IA when weighing 2,500 pounds or more.

(2) On shipments originating at Eldora, IA, when the total weight of all shipments, to all destinations, tendered and picked up at one time, from one shipper weighs 2,500 pounds or more. Bills of Lading covering separate shipments must be endorsed as part lot of at least 2,500 pounds tendered for pickup at one time.

NOTE 3—Provisions of this item will not apply from or to Grundy Center, IA on shipments weighing 2,500 pounds and over.

NOTE 4—Arbitrary rates will not apply from or to Hudson, IA, on shipments weighing 1,000 pounds or more or when freight charges are assessed on a minimum of 1,000 pounds. (160)

64

For explanation of abbreviations and reference marks see last page(s) of this tariff.

CHAPTER 6

COMBINATION THROUGH RATES

General

As we have already noted, through rates may be divided into three categories (1) local rates, those published to apply from origin to destination over the lines of one carrier, (2) joint rates, those published to apply through from origin to destination over the lines of two or more carriers and (3) combination through rates which are made up of two or more rates, either local or joint, applying via the route shipment moves.

In our study of combination rates one of the first questions to be resolved is whether the rate made by combining the separate factors is in reality a through rate in the legal sense of the term or whether each factor must be considered separately as applicable to an individual contract of carriage with the carrier or carriers over whose line or lines the rate applies.

We can see the legal nature of a combination through rate only by noting its reflection upon a given movement in contrast with a joint through rate. Therefore, we shall endeavor to make clear those features necessary to the establishment of combination through rates whereby they are put on a legal parity with joint through rates. This is accomplished principally by means of implied agreements for through carriage, as compared with definite agreements and concurrences in the case of joint through rates.

It must be remembered at the outset, however, that Rule 55(a) of Tariff Circular 20 provides:

> "When a rate, either local or joint, from point of origin to destination, has been established via a route, it becomes the only legal rate for through transportation via that route, whether it is greater or less than the aggregate of intermediate rates."

Tariff Circular No. M.F. 5 contains a similar rule reading as follows:

> "When a carrier or carriers establish a local or joint rate for application over any route from point of origin to destination such rate is the applicable rate of such carrier or carriers over the authorized route notwithstanding that it may be higher than the aggregate of intermediate rates over such route."

Before a combination through rate may be used it is necessary to determine whether or not there is a joint through rate published and if so the through joint rate must be applied except where the through tariff provides otherwise.

When carriers publish a joint through rate, such through rate is subject to the rules, regulations and routing restrictions (if any) contained in such tariff. In the absence of a joint through rate, however, it is necessary to check the lowest combination of rates in effect via the route over which the shipment is to·move and to particularly note whether or not a through route is in effect, since the carriers often limit the extent to which they will unite their respective systems (i.e., participate) in the formation of a single unit (route). If reasonable restrictions are noted, they are doubtless lawful and fully enforceable. If, however, the carrier's restrictions of their respective rates results in an unlawful situation, the matter can be taken up with the Interstate Commerce Commission which has power under Section 10704 of P.L. 95-473 to prescribe the rate (including a maximum or minimum rate, or both) to be followed.

Combination Through Rate Defined

Legally speaking, a combination through rate is one that applies over a through route which does not enjoy a joint through rate (one figure rate) from point of origin to destination, and which is made up by combining two or more separate factors.

For example, carrier X operates from A to B where it connects with carrier Y which runs from B to C as indicated below:

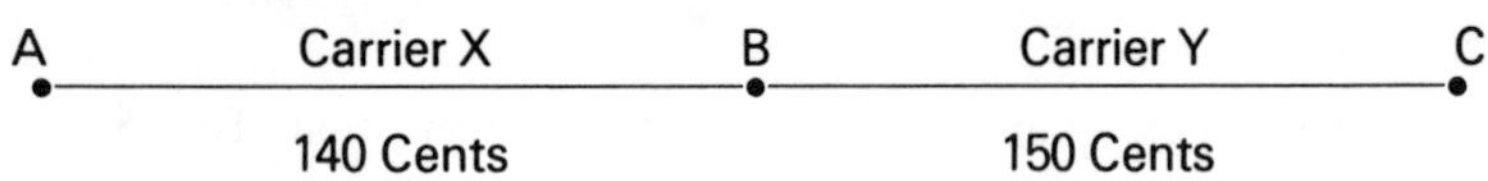

Assuming that a through route exists from A to C via carriers X and Y but that these carriers have not published a joint through rate to apply on movements between these two points, the rate from A to C would have to be computed by combining the rate from A to B of 140 cents with the rate from B to C of 150 cents making a through combination rate of 290 cents.

If carrier Y had a physical connection with carrier Z at point C for movements to point D and again no through rate was published the situation would be as shown in the following illustration with a through combination rate (three factors) of 400 cents.

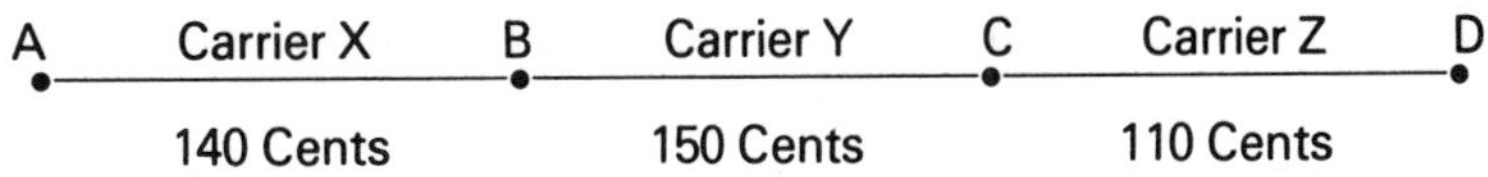

Assuming that a joint rate of 280 cents was subsequently established from A to C the rate from A to D would be a combination through rate (two factors, a joint and a local rate) of 390 cents.

In constructing a combination through rate there are no restrictions on the kinds of rates that may be used so long as they are legally established to apply via the route of movement. They may be two local rates as shown in the first illustration or they may be a joint rate and a local rate as illustrated above or they may be a local and a proportional or two proportionals or a joint rate and a proportional rate.

A rate that is published to equal the sum of the combination of separately published rates to and from a rate-break point with both factors being shown in the rate item is not a combination rate within the ordinary or usual meaning of this term. This principle was upheld by the Commission in *Weyl-Zuckerman & Co., v. A.T. & N.R.R.,* (210 I.C.C. 565), wherein they said:

> "The allegation of inapplicability is predicated upon the theory that a joint through class C rate takes precedence over a through commodity rate so published as to equal the sum of the combination of the separately published rates to and beyond a rate-break point. The fact that the through commodity rate is published in that manner does not make it a combination rate in the ordinary sense of the term. Where, as under the circumstances here present, a through commodity rate, by agreement of all participating carriers, has been published and filed in conformity with Section 6 of the Act, it is the applicable through rate, even though it is not in the form of a single-factor through rate."

Checking a Combination Rate

Individuals who have attained considerable proficiency in the use and interpretation of tariffs frequently encounter unwarranted difficulty in the checking of combination rates. Nearly always the root of the difficulty can be traced to the psychological stumbling block which the mere presence of a number of tariffs presents. In these comments we have tried to reduce the whole problem of combination rates to its very simplest terms so that the inexperienced traffic man will have a step by step guide showing the basic procedure.

The actual process of checking combination rates is no more difficult than checking two local rates. The following paragraphs describe the steps and mental processes in checking a typical combination rate.

If no previous experience has been had in the checking of rates on a particular movement there is no way to tell in advance the exact combination or the particular point over which the lowest combination rate may be made and it is, therefore, necessary to check all possible combinations. In order to do this, a good knowledge of the various rate breaking points throughout the country over which many combinations are made and the territorial application of the tariffs published by the carrier rate bureaus is extremely helpful. Some of the more important rate breaking points are the Mississippi River crossings, both upper and lower, the Ohio River crossings and the Virginia Cities shown below.

Upper Mississippi River Crossings:

East Burlington, IL	East Keokuk, IL	Quincy, IL
East Clinton, IL	East Louisiana, IL	Rock Island, IL
East Dubuque, IL	East St. Louis, IL	Savanna, IL
East Ft. Madison, IL	Keithsburg, IL	St. Louis, MO
East Hannibal, IL		

Lower Mississippi River Crossings:

Cairo, IL	Memphis, TN	Baton Rouge, LA
Thebes, IL	Vicksburg, MS	New Orleans, LA

Ohio River Crossings:

Cincinnati, OH Evansville, IN Jeffersonville, IN Louisville, KY

The Virginia Cities:

Alberta	Norfolk	Roanoke
Altavista	Petersburg	Suffolk
Lynchburg	Richmond	

In addition, there are other points which should be considered, i.e., Chicago, Peoria, Milwaukee, Minneapolis, Kansas City, Omaha, etc.

Disregarding any consideration of the need for some knowledge of the application of some of the tariffs within some of the various territories, and thinking only in terms of how combination are made and how to find them, let us consider the actual process of checking the rates. Our first thought should be, is there a through joint rate published between the points in question. If there is none, we must look for a combination of rates over the route of movement.

For the purpose of illustration as to the method of checking the lowest combination, let us consider a shipment of canned goods, carload, from a point A in Western Trunk Line Territory to a point B in Central Territory. Let us assume also that there is no through commodity rate in effect to the destination in question, but there is a through class rate and the tariff naming such through class rate carries an aggregate of intermediates clause allowing the use of the lower combination via all routes named in the through tariff.

Our first thought in considering a movement of this kind would naturally be the possibility of using either Chicago, Peoria or the Upper Mississippi River Crossings as possible rate-break points. Next, having located a tariff naming commodity rates from origin A east-bound to Mississippi River points etc., we find the following:

Commodity	From	To		
		Chicago	Peoria	Miss. River Crossings
Canned Goods, CL in boxes; Min. Wgt. 30,000	A	30	28	20

The next step is to locate a tariff naming rates from the above gateways to destination B. Let us assume we find a commodity tariff containing the following:

Commodity	To	From		
		Chicago	Peoria	Miss. River Crossings
Canned Goods, CL in boxes; Min. Wgt. 30,000	B	40	44	48

We now have the following result:

	Chicago	Peoria	Miss River Crossings
To	30	28	20
From	40	44	48
	70	72	68

From the foregoing it will be seen that the lowest combination would be the rate of 68 cents based on the combination over the Mississippi River Crossings.

Let us carry this one step further and assume that there are no commodity rates published beyond Chicago, Peoria and the Mississippi River Crossings and the only rates available were the class rates. It would then be necessary to check the class rate from Chicago, Peoria and each one of the river crossings referred to in the tariff from point A inasmuch as the class rate scale is predicated on distance and the rate would probably be different from the various crossings depending on their location with regard to the final destination.

The illustration used is a simple one which can be easily followed and while it is true that there are many much more complicated situations, nevertheless the principle and the procedure remains the same, and it all comes down to just one thing, namely, the checking of a rate from the point of origin to one or more intermediate points in the general direction of the destination and combining such rate or rates with a rate or rates from that particular point or points to the destination.

Essential Element of a Combination Through Rate

The essential element of a combination through rate is, that the carriers involved must have formed a through route via which the combination is made. It may be done by implication or agreement and, contrary to the essential element of a joint rate, need not be specifically and definitely concurred in but may be indicated by the circumstances and conditions surrounding the movement.

Through routes may be found to exist though no joint rate has ever been applicable to it. But the existence of a joint rate necessarily presupposes the existence of a through route; existence depends on the circumstances in each case; the test is whether the participating carriers hold themselves out as offering a through transportation service.

Where a through route exists the rate factors making up the combination through rate are those in effect at the time the shipment is accepted by the carrier at the point of origin. If there is no through route, each segment of the movement is subject to a separate contract and the combination rate would be made up of the separate factors in effect on the date of receipt by each of the individual carriers.

Legal Standing of a Combination Through Rate

As long as there is a through route formed between two or more carriers, the combination through rate applying over such through route holds the same standing in law as a joint through rate. Both are the results of agreements between carriers to merge into a single system from point of origin to destination, the only difference being in the form in which such joint agreement is indicated. In the case of a joint through rate the agreement takes the form of a specific concurrence and the publishing of a *one figure* rate in the tariff, while in the case of a combination through rate it is indicated by the circumstances which control the formation of a through route; i.e., the collection and division of freight charges, use of proportional rate to or from junction or basing points, issuance and subsequent recognition of through bill of lading, etc.

Soon after its establishment, the Commission passed upon the principle of a combination through rate being on a legal parity with a joint through rate, despite the difference in their appearance and method of publication, when, in the case of *Milwaukee Chamber of Commerce v. Flint & Pere Marquette Railroad Company, et al.*, (2 I.C.C. 393, 398), it said:

> "It was wholly immaterial whether such a rate so made was 'quoted' as a through rate or not. The through bills of lading, the fixed total rate, the waybill, the expense bill, the course of business between the parties by which the contract of shipment was performed from the point of origin to destination made it a through rate. Names are nothing in such a transaction. The law looks at the elements and substance of the transaction itself. The through bill of lading issued by the initial carrier in which it was shown what the rate was, and that the carriers would transport the freight from Minneapolis to Milwaukee and east to destination, was a contract with the shipper for a through rate. The proposal of the initial carrier was accepted by the defendants and their connecting lines when they took the freight at

> Milwaukee and transported it to New York upon their agreed rate of 25½¢ per 100 pounds, and it then became an executed contract, and performed according to the well known and published course of business of these carriers. The shipper had contracted for the through rate, and had received it. It would be a mockery in terms to call such a transaction any other than a through rate. ***
>
> "A rate is nonetheless a through rate when freight is shipped upon a through bill of lading from the point of origin to destination, accompanied by a waybill, showing the route over which it is to go, with percentages of all the other lines set forth on the waybill, because the initial carrier charges its local rate as part of the total rate, and the remaining lines charge an agreed rate made by percentages. It may occur where the freight is shipped under a through bill of lading from the point of origin to the final destination, and has to pass over ten or a dozen different lines of railroad, and several of these, or, for that matter, each of these roads may charge its local rate, and still the total rate is a through rate."

Throughout the Commission's decision we see an unwavering adherence to this principle of law. *In the Matter of Through Routes and Through Rates,* (12 I.C.C. 163, 172), it stated:

> "A combination through rate is as binding, definite and absolute as a joint through rate; and all of the conditions, regulation, and privileges obtaining as to any factor in such combination rate for through shipment at the time of initial shipment upon such combination through rate must be adhered to and cannot be varied as to that shipment during the period of transportation of such shipment to its final destination."

In the case of *Board of Trade of Kansas City, Mo., et al, v. A.T. & S.F. Ry. Co., et al.,* (69 I.C.C. 185, 188), and in *Larabee Flour Mills Corp. et al., v. A.T. & S.F. Ry. Co., et al.,* (148 I.C.C. 5, 9), we find the following very forceful reiteration of its earlier decision in this respect:

> "A through rate, however made up, whether joint or in some form of combination, still remains a through rate and necessarily must be that which was in effect at the time of shipment from the point of origin."

Also, in Rule 55(b) of the Commission's Tariff Circular No. 20, we observe a crystallization of this principle into a specific recognition of a method of constructing combination through rates on through shipments where no joint through rate exists, as follows:

> "(b) If no rate is named from point of origin to destination of a shipment via the route of movement, the lowest combination of rates applicable via the route of movement is the legal rate.
>
> "A combination rate for a through shipment must be treated as a unit from point of origin to final destination, and the rate applied must be the

combination of the factors in effect on the date the shipment was accepted for transportation at point of origin. All the conditions applicable to each factor in such combination rate for through shipment in effect on the date the shipment was accepted for transportation at point of origin must be observed and cannot be varied as to that shipment during the period of its transportation to final destination."

Here, then, we have a recognized principle of law that it is not the appearance or the manner in which it is published that is the controlling factor in determining the existence of a through rate but it is the real substance underlying that appearance which is considered. If, therefore, there is an arrangement for continuous carriage between carriers and, consequently, a through route has been formed between them, the rate applying over such through route is a through rate. If the arrangement between the carriers is the result of an expressed agreement evidenced by proper concurrence in a duly authorized tariff, the rate applying over the through route is a joint through rate. If it is not definitely evidenced in this manner, but the circumstances and conditions surrounding the movement do nevertheless indicate the existence of a through route, the through rate applicable over such through route is a combination through rate; and its separate factors apply *as of date of shipment from point of origin,* just as though they actually had been merged into one unit by a joint agreement and published concurrence.

In the event that there is no through route formed between the carriers, there would be no agreement for a combination through rate and the factors entering into such combination rate would be a "combination of locals" and would apply *as of date of receipt of the shipment by the separate carriers.*

Attitude of Commission in Respect to Combination Through Rate Versus Joint Through Rate

It is an accepted principle of rate making that when a joint rate is established between two points, such joint rate is, because of the reduction in the number of terminal expenses involved, usually made lower than the combination of local rates applying between the same points. As a combination through rate, therefore, invariably results in the assessment of charges that are higher than would obtain under a joint through rate, it is quite logical to inquire why the Commission has not required the carriers to publish reduced joint through rates to be applied on through business instead of permitting them to assess their higher combination of separate locals.

In connection with this feature, the Commission, in the case of *Intermediate Rate Association v. Director General, Aberdeen & Rockfish Railroad Company, et al.*, (61 I.C.C. 226, 246), stated:

> "When distances are relatively great, and when transfer at rate-breaking points is not attended by unusual costs, the combination basis, using local rates, ordinarily is abnormal and unscientific and often discriminatory. The railroads should be regarded more and more as one national system, and the time may not be far distant when we should proceed to the establishment of joint through class and commodity rates, lower than the combinations of locals, between practically all points in the country. We have generally recognized that through rates should be less than the combinations, but prompted chiefly by considerations of paramount public interest, growing out of the revenue conditions of certain carriers, we have refrained from and even declined absolute condemnation of combinations."

This citation shows that, because of the then unsettled financial conditions of the carriers, the Commission deemed it unwise to lower the level of their income by requiring a general or blanket establishment of joint through rates lower than the sum of the locals. This should not be construed to mean that the Commission will not in individual cases, if the circumstances are sufficient to warrant it, grant relief of this nature. The Commission has since fulfilled its prophecy by prescribing joint through rates by rail between all territories in the United States as found in the Consolidated Southwestern, Western Trunk Line, Eastern and Southern class rate investigations, and subsequently in *Docket No. 28300, Class Rate Investigation, 1939,* (262 I.C.C. 447). and *Class Rates, Mountain-Pacific Territory Class Rates, Transcontinental Rail, 1950,* (296 I.C.C. 555).

In *A.C.L. R.R. v. Southern Ry.*, (321 I.C.C. 314), the Commission ruled:

> "Contention that combination rates applicable on cotton from Mississippi Valley origins to southern territory destinations over routes restricted against application of revised joint rates by defendant's tariff routing provisions are unreasonable in violation of Section 1(4) and (5), because they are higher than the revised joint rates, would have merit only if the restricted routes were shown to be through routes. However, all of the restricted routes have been closed to that traffic since effective date of defendant's revised routing provisions, and mere existence of higher combinations over those routes does not violate Section 1(4) and (5)."

And in *Southwest Fabricating & Welding Co., v. Alton & S. R.R.*, (302 I.C.C. 440), the Commission stated:

"There were no joint through rates from any of the primary origins to any of the ultimate destinations which could be applied with transit at designated points. Hence, applicable rates were the lowest combinations over the route of movement."

COMBINATION OF PROPORTIONALS

General

A proportional rate, as its name implies, represents only a part of a through rate for a transportation service. Therefore, in order that it may complete the purpose for which it was created; i.e., the fusion of its existence into the unity of a combination through rate, it is necessary to understand the principles governing the use and application of proportional rates in combination with other rates in arriving at the lowest total through charges.

To locate and utilize any particular proportional rate in connection with a specific situation is purely a matter of tariff application and interpretation. In fact, the Interstate Commerce Commission states in Tariff Circular No. 20 and M.F. No. 5, that "Tariffs containing proportional rates must clearly and definitely show the application thereof." The rule also goes on to state that "if the application is not restricted, such proportional rates will be usable in connection with any other applicable rates from or to the proportional rate point. It will not be permissible for a tariff to state that a proportional rate applies from (or to) points from (or to) which no through rates are published. Such a provision is not sufficiently definite to restrict the application of the rate. If a proportional rate is intended for use on traffic destined to a restricted territory, such territory should be clearly defined. For example, a tariff naming a proportional rate to St. Louis intended for use on traffic destined to Kansas should state that the rates apply on traffic destined to Kansas only."

The propriety and lawfulness of proportional rates to the points of transfer which are less than local rates to that point have frequently been affirmed by the Commission, and are sanctioned by considerations of public policy. The idea of differing proportional rates for the same service depending upon the origin or destination of the traffic did not originate with the Commission. It originated with and has been used by carriers to a great extent.

In *Refund Provisions, Lake Cargo Coal,* (220 I.C.C. 659), the Commission said:

> "Proportional rates are in the nature of divisions of through rates which may vary depending on the origin or destination of the traffic without necessarily violating any provision of the Act. Such variations, when properly applied, are consistent with sound principles of rate making."

Again, in *American Barge Line v. A.G.S. R.R.*, (303 I.C.C. 463), the Commission held:

> "Proportional rates have been recognized as important parts of the ratemaking mechanism necessary to maintain interrelated and lawful rates on through traffic, such as grain, particularly when transit services are authorized in the tariffs. They are but parts of through rates, and are capable of use only as such."

Nature and Application of a Proportional Rate

A proportional rate is a proportion or part of a through rate, and is employed principally as a means of naming through rates lower than the sum of the intermediate local rates where no joint through rate has been established.

In *Hocking Valley Railway Company v. Lackawanna Coal & Lumber Company*, (224 Fed. Rep. 930, 931), the Circuit Court expressed itself quite clearly along these lines, in the following language:

> "A 'proportional rate,' as the term implies, is simply a part of a through rate. It is the share of the aggregate charge from origin to destination which one or more of the carriers accepts for performing a definite portion of the whole transportation service. It is a matter of common knowledge that through rates are generally less than the sum of intermediate local rates; and when all the participating carriers do not join in establishing the through rates, it is a common practice for one or more of them to name proportional rates up to some point of connection with another carrier which completes or continues the transportation. The propriety and lawfulness of proportional rates to the point of transfer which are less than local rates to that point have frequently been affirmed by the Interstate Commerce Commission, and are sanctioned by consideration of public policy."

Since a proportional rate between two points must be linked with another rate before it may be used, the question naturally arises as to what the relationship is between a proportional rate and the contemporaneous local rate between the same points.

The following expression of the Commission, as contained in the case of *Indian Refining Co., Inc., v. L. & N. R.R. Co., et al.*, (112 I.C.C.

732, 735, 736), shows clearly the distinction that exists between local and proportional rates:

> "The primary purpose of a local rate is to apply to local shipments of the commodity which it covers from and to the points specified, although in the absence of joint through rates or proportional rates of the same character between the same points, it applies on through traffic as well as local traffic. Our tariff rules do not authorize or permit the publication of conflicting rates. They provide that where a commodity rate is established between two points it removes the application of the class rate between the same points without requiring the cancellation of the class rate. Our rules also authorize the publication of proportional rates and are silent as to necessity for cancelling the local rates between the same points for application on through traffic. Since the rules do not provide that proportional rates conflict with local rates, it follows that the establishment of a proportional commodity rate removes the application on through taffic of a local commodity rate on the same article between the same points."

In *Koppers Co. v. C.& O. R.R.,* (303 I.C.C. 383), the Commission said:

> "The rate structure of the railroads consists principally of systems of single-line and joint-line rates, with a per-mile progression diminishing more or less with increasing distance, and hence, on joint line traffic, considerably lower than the combination of local rates. When joint rates have not been desirable, or have not been possible, as in the case of rail-water routes where the water transportation is not subject to the Commission's jurisdiction, substantially the same result has been accomplished by the establishment of so-called proportional rates for the rail haul or for a portion of the rail haul."

Proportional Rates Must Be Filed with Interstate Commerce Commission

Section 6, paragraph 1, of the former Interstate Commerce Act provided that:

> "If no joint rate over the through route has been established, the several carriers in such route shall file, print and keep open to public inspection as aforesaid, the *separately established* rates, fares, and charges applied to the through transportation."

Parts II, III and IV did not in specific terms contain a similar requirement, but stated that carriers must file and keep open to public inspection tariffs showing all the rates and charges for transportation and services. A similar provision is contained in Section 10762 of P.L. 95-473.

Since a proportional rate is a part of the through rate applying over a through route, and, therefore, comes within the above provisions of the law providing for publication and filing of "the separately established rates applied to the through transportation," a common carrier subject to the Act must, before it can engage in interstate commerce under such rates, file its proportional tariffs with the Interstate Commerce Commission.

In passing upon this feature, the Commission, in the case of *Greater Des Moines Committee, Inc. v. Director General, as Agent C.M. & St.P. Railway Company, et al.*, (60 I.C.C. 403, 407), stated:

> "Being separate parts of through rates, proportional rates, in order to be lawful, must be established, within the meaning of Section 6, by publication and filing with us."

It is understood, of course, that this requirement applies only to *interstate* business. If the shipment is not interstate but transportation takes place wholly within the state, it does not come within the jurisdiction of the Interstate Commerce Commission, and any proportional rate properly applicable to such intrastate movement need not necessarily be filed with the Interstate Commerce Commission.

This should not be construed, however, to mean that the movement covered by the proportional rate alone is all that need be considered in determining whether the proportional rate should be filed with the Interstate Commerce Commission. As the proportional rate is a part of the through rate, it is apparent that the entire movement covered by such through route must be considered. If the shipment originated in one state and terminated in another state, the through rate covering such shipment would be an interstate rate. It would, therefore, not be lawful to attempt to break the proportional rate away from the balance of the through interstate rate and claim that such proportional rate was not subject to the Interstate Commerce Commission's supervision because it, in itself, applied only between points within the same state.

Local Rates Published as Proportional Rates

In connection with the publication and filing of proportional rates by the carriers, it will be found that in many instances a carrier will publish local rates in a tariff and in addition show on the title page of such tariff that it also contains proportional rates. Upon referring to the contents of the tariff, however, it will be found that no proportional rates, *as such*, are contained in the publication. While, at first thought, it might appear as though this were an error in tariff compilation, when consideration is given to the fact that the local rates named in such tariff are intended to be used in combination with other rates, or as *portions* of through rates on business handled in connection with other roads, the apparent discrepancy becomes recognized as a very fine point of tariff construction and a rate technicality that is not generally understood. In explanation of this feature, you will note, the law requires the individual carriers to publish and file the "separately established rates, fares, and charges applied to the through transportation." By publishing and filing their local rates with the Commission and at the same time indicating that they may be used as proportional rates, the carriers impart to their tariffs a double duty; that is to say, they meet the requirements of the law in respect to the filing of their local rates on interstate business and at the same time comply with the requirement of the Act which calls for the filing of the "separately published" rates, when such local rates are used as portions of the through rates in connection with through movements.

The Commission, in originally speaking of the difference which exists between the local nature and the proportional nature of a rate published in this manner, made the following statement in the case of *Interior Iowa Cities Case,* (28 I.C.C. 64), and held:

> "Proportional rates are necessarily parts of through rates and differ from local rates used as part of through rates in that before the proportional rate may be attacked at all there must be an allegation that the through rate is unreasonable because of the unreasonableness of the particular proportional rate; whereas local rates, as such, may be attacked separately when used separately.
>
> "A shipper has no legal grievance with respect to this through traffic unless compelled to pay excessive charges for through service. If the through charges are lawful in the sense that they are reasonable charges for the through service, a shipper can not predicate unlawfulness of one of the component parts of the through charges by alleging that it is excessive compensation to that carrier for that part of the through service. He pays for completed service, and it is no concern of his how the through charges are divided among the carriers, whether by agreement or by published proportionals, so long as the through charges for the through carriage are reasonable. The futility of a finding that a proportional rate is excessive, in the absence of a finding also that the through charges are excessive, is clear, for by so much as we may reduce the proportional rate, of one carrier another carrier in the route may increase his proportional rate, thus leaving the through charges unaffected by our order."

And in *Stevens Grocer Company, et al, v. St.L.I.M. & S. Railway Company, et al.,* (42 I.C.C. 396, 398):

"Local rates, however, when used as parts of through rates partake of the nature of proportional rates, and may be regarded as, in effect, local and proportional rates. So far as they are strictly local rates they may, of course, be brought in question without questioning the propriety of any other rate; insofar as they are used in through transportation they should be treated as other proportional rates and may not be considered unless the through rate be attacked as a whole."

It will be observed from these citations that it is not the apperance or the manner in which a rate is published that determines its substance. In other words, while a proportional rate may, as a matter of fact, be published in the same tariff as a local rate, present the same appearance as the local rate, or even be the identical figure as the local rate, there still exists a wide legal difference between them not compatible with their physical similarity. This legal difference concerns itself with the manner in which the rate is applied, and, in the final analysis, adheres to the principles above indicated; i.e., that where a local rate, as such, is applied, it is a complete unit and may therefore be brought in issue; but where it is used as a proportional rate in connection with a through movement it loses its individuality and becomes one of the factors entering into the making of the through rate. As it is then nothing but a part of the through rate, its unity is destroyed, and it is the through rate as a unit which is given paramount consideration, the proportional rate in such cases being considered only insofar as it, as a factor, may affect the reasonableness or unreasonableness of the through rate.

The following expression of the Commission, as contained in the case of the *Stevens Grocer Company, v. St.L.I.M. & S. Railway Company,* previously referred to, is further indicative of this principle:

"It is important that the true rule be definitely announced and that a uniform policy be established under which parties complainant and defendant may understand what is required. We now lay down the rule, which for the future will be strictly adhered to, that when a complaint involves charges applicable to a through shipment the through rate or charge must be brought in issue and the participating carriers must be made defendants. When the through rate or charge is made up of separately established rates or charges, applicable to the through business, the through rate or charge must be attacked as violative of the Act, although the violation may be believed to be occasioned by a particular factor or factors thereof; in such case the complainant should be prepared at the hearing to prove the unlawfulness of the through rate itself and that this is due to a particular factor or factors."

It is to be noted that this rule as laid down by the Commission applies to that class of cases where the charges resulting from the application of a combination through rate are alleged to be unreasonable. Where such charges are not brought into issue, the Commission holds that they have the power to pass upon the lawfulness of a proportional or reshipping rate. This feature was clearly brought out in *Cairo Board of Trade v. C.C.C. & St.L. Ry. Co.,* (46 I.C.C. 343, 350), as follows:

> "The rule is stated in the *Stevens Grocer Co.* case more broadly than it should be. In determining whether or not a complainant has been damaged by the exaction of unreasonable or unduly preferential reshipping rates, the total through charges paid from a point of origin must be considered. But this does not hold true of a determination of the reasonableness or justness of the reshipping rate itself. Reshipping rates are not merely divisions of through rates, but are separately established rates generally published by carriers other than those engaged in the inbound movement and without the concurrence of the latter; and the point of reshipment is a rate-breaking point. A change in the reshipping rates, even though it may affect the through charges, will have no effect upon the inbound rates. The inbound carriers have a right to secure reasonable compensation for their part of the haul by reasonable inbound rates. The reasonableness of such inbound rates is in no manner contingent upon reshipping rates. Furthermore, inbound rates used in connection with reshipping rates generally serve also as local rates. Hence they are subject to review independently of the outbound rates. An excessive reshipping rate might produce a reasonable through charge in connection with an unduly low inbound rate and vice versa. It can not properly be argued that a proposal to increase unremunerative reshipping rates could be denied upon the ground that the through charge composed of an excessive inbound rate and the unremunerative reshipping rate is just and reasonable. The converse must also be true, namely, that shippers may not upon the grounds be denied relief from unreasonable or unduly prejudicial reshipping rates. This is also true as to proportional rates that are applicable to shipments going or from beyond and which are not limited as applying only on shipments from or to designated points or territory. Each of such rates must be judged upon its individual merits."

The Commission in contrast to its findings in the *Interior Iowa Case* and *Stevens Grocer Company,* supra, further dealt with this subject in *Fraser-Smith Co. v. Grand Trunk Western Ry.,* (185 I.C.C. 57, 58), in the following language:

> "In the *Cairo Case* we considered our authority to pass on the unlawfulness of separate components of combination rates where other factors of the combinations were not before us. We found that in such a case we could consider the lawfulness of the separate components for the future, but that in determining whether a complainant had been damaged by the exaction of unlawful components of through rates, we must consider the total through charges paid from point of origin."

Reasonableness of Proportional Rates Cannot Be Measured by Corresponding Local Rates

It should not be inferred, from what has been said in regard to a local rate being used as a proportional rate, that a local rate is the only kind of a proportional rate. The more common form of proportional rate is one which is published entirely separate from a local rate and which, as a rule, is made lower than the local rates between the same points. In passing upon the propriety of proportional rates which are lower than the corresponding local rates, the Commission, in the case of *Capital Grain & Feed Co. v. I.C. R.R. Co.,* (118 I.C.C. 732, 743), stated:

> "Proportional rates ordinarily are lower than local rates from and to the same points."

It has also been held in *Chicago & Wisconsin Proportional Rates, supra,* that it is a generally accepted principle that traffic is entitled to proportional rates lower than local rates only when the movements for which the proportional rates apply are continuous from and to points beyond the proportional rate point.

This does not mean, however, that the reasonableness or unreasonableness of a proportional rate can be determined by the fact that it is or is not lower than the corresponding local rate. While it is in the nature of a *division* of the through rate and is therefore usually lower than the local rate, in some cases it may be equal to or even exceed the local rate between the same points, as will be evidenced by the following finding of the Commission as contained in the case of *Boney & Harper Milling Company v. Atlantic Coast Line Railroad Company, et al.,* (28 I.C.C. 383, 388):

> "We have frequently said that a proportional rate applying on through traffic might well be less than the corresponding local rate, but we have not said that such proportional rate must be, or in every case, should be, less. There may be, and are, many instances in which full local rates are applied as proportions in the construction of through rates, and the rate so established in this case can only be condemned upon a satisfactory showing that it is unreasonable, unduly discriminatory, or otherwise in contravention of law."

Therefore, in summing up, it is well settled that a proportional rate, (1) may be the same as a local rate or it may be higher or lower depending upon circumstances; (2) it may be attacked separately as an allegation of unreasonableness for the future, but, when reparation is involved the through rate of which it is a part must be considered in establishing damage in the past.

Chapter 7

AGGREGATE OF INTERMEDIATE RATES

General

In earlier chapters we dealt with the establishing of joint through rates and the various combinations of rates that may go into the making of a through rate where there is no specific joint through rate published from point of origin to destination or where no specific manner of constructing the combination through rate is prescribed.

In pursuing the subject of through rates further, we arrive at a study of those instances where there is a joint through rate specifically published and on file with the Interstate Commerce Commission, and which is higher than the combination of the separately established intermediate rates also on file with the Commission and which are in effect over the same through route between the same points. For example, let us say that we have a specific joint through rate of 95 cents per 100 pounds in effect between points A and C and that over the same route in the same direction there exists a lower combination of rates amounting to 90 cents, as indicated by the following diagram:

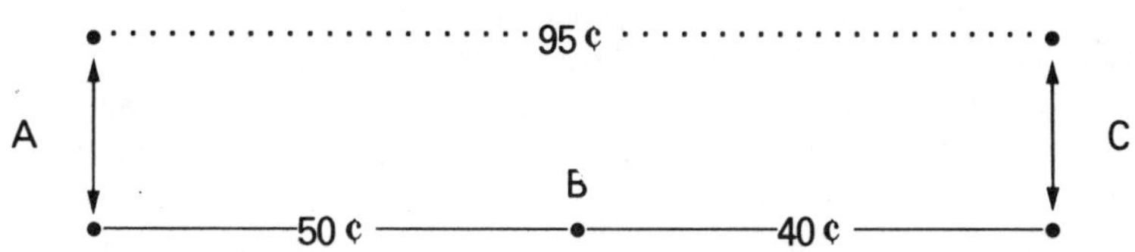

Here, then, we have a combination through rate of 90 cents, which is lower than the published joint through rate of 95 cents.

As previously stated, a joint through rate is usually lower than the aggregate of the intermediate rates, although, in some instances, such published through rate may be the same as the combination rate. According to the wording of Section 10726 of P.L. 95-473, it is unlawful for a through rate to exceed the aggregate of the intermediate rates. It is this feature and the authority of the Commission to permit departure from this provision which we will discuss in this chapter.

Unlawfulness of Joint Through Rate Exceeding Combination of Locals

The aggregate of intermediate rates provision of Section 10726 of P.L. 95-473 is based upon the unreasonableness of a situation which would permit a carrier to charge more for performing a through transportation service which included but two terminal services, as contrasted with four or more such expenses entering into a lower combination of locals, and which would also yield to it a greater ratio of earnings on the longer haul than on the separate shorter hauls, despite the fact that the cost of performing the through service is proportionately lower than that of handling two or more shorter movements.

In speaking of this feature, the Commission in the case of *G.L. Jubitz, Assignee, v. Southern Pacific Railway Company, et al.*, (27 I.C.C. 44, 45), stated:

> "Ordinarily, as the length of haul increases, the rate per ton mile decreases; and, generally speaking, a joint through rate should be somewhat less than the combination upon an intermediate point, for the reason that in the through haul there are but two terminal expenses."

It is self evident that if a rate between two points is reasonable and the traffic handled thereunder bears its fair share of the burden of transportation and is reasonably compensatory to the carrier, and the same is true of a second intermediate rate, these two intermediate rates when combined, afford the carrier a reasonable combination rate for the through movement. In other words, if, according to the diagram on page 107, the carrier can haul goods in interstate movement from "A" to "B" at a profit and from "B" to "C" at a profit as separate movements, there can be no justification for combining the rates covering these movements and then adding on something additional, on traffic moving through from "A" to "C." This is particularly evident when we realize that the expense of moving through business is less than the aggregate expense of the shorter separate movements, and that, therefore, the through rate should be at least as low or lower than the combination.

Section 10726

Section 10726 of P.L. 95-473 is a legislative recognition of this principle of rate making. This section reads in part as follows:

> "A carrier providing transportation subject to the jurisdiction of the Interstate Commerce Commission under subchapter I or III of Chapter 105

of this title (except an express carrier) may not charge or receive more compensation for the transportation of property of the same kind or of passengers—

* * *

(B) under a through rate than under the total of the intermediate rates it may charge under this chapter."

It will be noted that this so-called aggregate of intermediates clause is applicable only to subchapter I carriers (railroads) and subchapter III carriers (water carriers). There is no comparable provision applicable to motor carriers.

Tariff Circular Rule 56

Prior to the amendment of June 18, 1910, the Act contained no definite restriction against the charging of more as a through rate than the aggregate of the intermediates. However, the Interstate Commerce Commission and the courts, both State and Federal, readily perceived the unreasonableness of a situation of this nature, and consistently held that a through rate which exceeded the aggregate of the intermediate rates over the same line in the same direction was prima facie unreasonable and that it rested upon the carrier to prove the justification thereof, if any existed. The following ruling of the Commission, which was issued on December 31, 1906, and later incorporated in Tariff Circular 14-A as Rule No. 43, and which is now carried in the Commission's current tariff circular as Rule No. 56(a), is indicative of its attitude in this respect:

"(a) Section 4 of the Act, as amended, prohibits the charging of any greater compensation as a through rate than the aggregate of the intermediate rates that are subject to the Act. The Commission has frequently held that through rates which are in excess of the sum of the intermediate rates between the same points via the same route are prima facie unreasonable.

"Many informal complaints are received in connection with regularly established through rates which are in excess of the sum of the intermediate rates between the same points. The Commission has no authority to change or fix a rate except after full hearing. It is believed to be proper for the Commission to say that if called upon to formally pass upon a case of this nature it would be its policy to consider a rate which is higher than the aggregate of the intermediate rates between the same points via the same route as prima facie unreasonable and that the burden of proof would be upon the carrier to defend such unreasonable rate."

Among the early formal cases in which the Commission has held that a through rate higher than the aggregate of the intermediates was prima facie unreasonable are the following: *Windsor Turned Goods Co. v. C. & O. Ry. Co.*, (18 I.C.C. 162); *Milburn Wagon Co. v. L.S. & M.S. Ry. Co.*, (18 I.C.C. 144); *Smith Mfg. Co. v. C.M. & G. Ry. Co.*, (16 I.C.C. 447); *Mich. Buggy Co. v. G.R. & I. Ry. Co.*, (15 I.C.C. 297); *Hope Cotton Oil Co. v. T. & P. Ry. Co.*, (112 I.C.C. 265).

As further recognition of the unreasonableness of a through rate in excess of the aggregate of the intermediates and of the injustice which would result if some means were not given to speedily relieve a situation of this kind, the Commission, in paragraph (b) of Rule 56, waived the requirement of thirty days' statutory notice, in order to permit the carriers, upon one day's notice, to reduce such higher through rate to the sum of the intermediates. This rule reads in part as follows:

> "Where a rate is in effect by a given route from point of origin to destination which is higher than the aggregate of intermediate rates from and to the same points, by the same or another route, such higher rate may, on not less than one day's notice to the public and the Commission, *be reduced to the actual aggregate of such intermediate rates.*"

The Commission also provided in paragraph (c) of Rule 56:

> "In order to facilitate the publication of rates which will be in accord with the aggregate of intermediates provision of the fourth section of the Act, the following rule may be incorporated in tariffs:
>
> > "Carriers have endeavored to publish herein rates which do not exceed the aggregate of the intermediate rates between points between which there is an actual movement of traffic, but if there should be in this tariff any rate which is in excess of the intermediates, or if through subsequent change in an intermediate factor any rate in this tariff becomes higher than the aggregate of intermediates in violation of the provisions of the fourth section of the Act, carriers will reduce such rate to the aggregate of the intermediate rates on one day's notice under authority of Rule 56 of Interstate Commerce Commission Tariff Circular No. 20, on any commodity between points between which there is a movement or a prospective movement of that commodity. The publication of such reduced rate will be made within thirty days after such unlawful rate comes to carrier's notice.
> >
> > "Carriers, parties to this tariff, whose rate over the route of movement is higher than the aggregate of the intermediates over that route, further agree that on any shipment on which the higher rate named in this tariff for that route has been charged, application will be made promptly to the Interstate Commerce Commission for authority to award reparation on the basis of the aggregate of intermediates in effect on date of shipment. (See note.)

"Note—Carriers or shippers who discover combinations which result in lower charges than the rates named herein, should promptly report such cases to the publishing agent of this tariff, showing the through rate and the item or page where it is found, together with the separate factors which make up the combination, giving tariff reference by item or page, where possible, for each."

Further in this administrative ruling, 56(c), the Commission warns the carriers:

"(2) In placing this rule in tariffs cariers must strictly adhere to the wording of the rule as no modification thereof will be permitted.

"(3) The failure to publish and file reduced rates as provided in this part, within 30 days from the date that said rates are brought to the attention of the carriers party thereto, or any of them, or their agents, will be considered by the Commission as sufficient ground for the issuance of an order prohibiting its use in connection with such carrier or carriers. A promise to publish certain rates, when published in a tariff, becomes the rule of the carriers parties to the tariff and therefor when a carrier or agent has been called upon to reduce rates under authority of the above rule it will not be necessary for such carrier or agent to secure any additional authority from the carriers parties to the tariff for the publication of the reduced rates and any delay on that account may cause carriers to incur the penalties provided for violations of the fourth and sixth sections of the Act in addition to losing the right to use the rule in tariffs."

Since October 1, 1962, railroads have been allowed to publish the following rule in their tariffs, in lieu of Rule 56:

"If on any shipment an aggregate-of-intermediate local, joint and/or proportional interstate rate constructed via a route over which the through rate published in this tariff is applicable produces a lower charge than the through rate, such aggregate of rates will apply via all routes authorized in this tariff and the through rate has no application to any such shipment via any of those routes."

Whenever this provision appears in a rail tariff it has the effect of alternating the through rate in that tariff on a given shipment with the combination rate over any route authorized in the tariff naming the through rate. Thus, if a combination is found which is lower than the single factor through rate, the combination is applicable and it is not necessary for the carrier to specifically publish a reduced single factor rate reflecting the combination, although the carriers may very well elect to do so.

An exception to the above open aggregate of intermediates provision is contained in Trunk Line Tariff Bureau, East-South Class Rate Tariff

1008. Interstate Commerce Commission Order 52472 of January 15, 1952 authorized the rail carriers to depart from Rules 4(h), 4(i) and 7(b) of Tariff Circular 20 and publish an aggregate of intermediates rule permitting the application of a combination rate lower than the through but only when such lower combination applied via the route shipment moved.

Unless one of these rules is published in the tariff, the carrier must assess the published single factor through rate even though both shipper and carrier agree that a lower combination exists. Rule 55(a) of Tariff Circular 20 states:

> "When a rate, whether local or joint, from point of origin to destination, has been established via a route, it becomes the only legal rate for through transportation via that route, whether it is greater or less than the aggregate of intermediate rates."

Although a single factor through rate higher than a combination is subject to attack on various grounds before the Commission, Rule 55(a) definitely establishes it as being the applicable rate, in the absence of an aggregate of intermediates provision permitting the use of the lower combination. When no such provision is contained in the through tariff the shipper's only recourse is to request the carrier or carriers involved to reduce the rate to the level of the lower aggregate of intermediates on one day's notice under authority of Rule 56 of the tariff circular. After this has been done, the carrier handling the matter should be requested to make application to the Commission on the Special Reparation Docket for an order awarding reparation based on the difference between the published through rate and the lower combination. If the Commission is convinced that the intermediate rates are not unreasonably low, it will normally allow the carrier to make reparation.

Exceptional Circumstances in Justification of Higher Through Rate

There are some instances where exceptional circumstances justify the existence of a through rate higher than the aggregate of the intermediate rates subject to the Act. In *Armour & S.F. Ry.*, (291 I.C.C. 223), the Commission described combination rates under the aggregate of intermediate rule as follows:

> "Rates under an aggregate of intermediates rules are not combination rates as that term is ordinarily used. The aggregate rule is an alternative rate-making provision published for the purpose of protecting the lowest

aggregate rate made over any published through routes for application over all routes subject to rates from and to the same points without discrimination. The rule was intended to apply on traffic which prior to establishment of the through rates moved on combination rates, and which continue to move on that basis with the privilege of liberalized routings."

The definition in the Armour case is an interesting concept, however, the fact remains that the through rate arrived at under an open aggregate-of-intermediates rule consists of two factors which are combined to produce a through rate. The important thing to remember is that most inter-territorial class rate tariffs contain such an aggregate-of-intermediates clause and it is frequently possible to find a combination of rates that will produce a lower charge and which may be applied via any route named in the through tariff.

The following diagram shows the application of an open aggregate-of-intermediates rule. A shipment moving from Kansas City, Mo. to New York, N.Y. was routed via St. Louis, over which route a through rate of 250 cents applied. The rate from Kansas City to Chicago is 80 cents and from Chicago to New York the rate is 150 cents, producing a through combination rate of 230 cents.

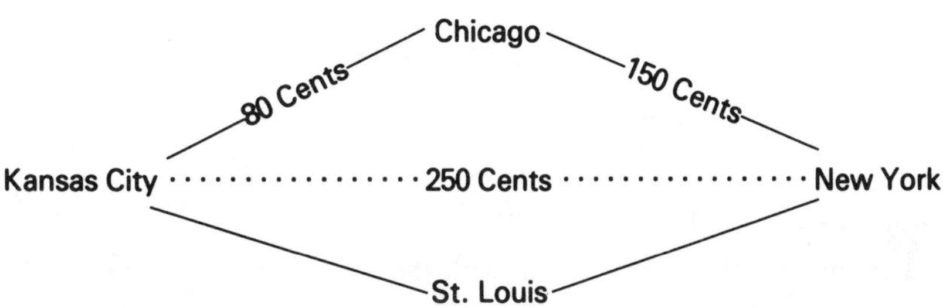

Under the open aggregate-of-intermediates clause the combination via Chicago may be applied despite the fact that the shipment was routed and moved via St. Louis.

In some instances the Commission has held that a "joint rate in excess of the aggregate of intermediate rates over the same through route presumably is unreasonable to the extent that it exceeds the aggregate of the intermediate rates, but that the presumption is not conclusive and may be rebutted." See *O.W. Slane Glass Co. v. Va. & C.W. Ry. Co.,* (39 I.C.C. 586); *Humphreys-Godwin Co. v. Y. & M.V.R. Co.,* (31 I.C.C. 25); also *Von Platen Box Co. v. Ann Arbor R. Co.,* (214 I.C.C. 432, 436).

That it is possible for a carrier to prove that sufficient reason exists for the charging of such (presumably) abnormal rate was recognized by the U.S. Supreme Court in the case of *Davis, as Agent, etc., v. Portland Seed Co.,* (264 U.S. 403, 425). In the words of the Court, "If a lower rate published without authority becomes the maximum which may be charged from any intermediate point, mistakes in schedules (and they are inevitable) may be disastrous." In cases brought before it involving this feature, however, the Commission has almost invariably, *upon proper proof of damages,* awarded the complainant reparation for the difference paid between the lower aggregate of intermediates and the published through rate.

An illustration of "exceptional circumstances" is found in *Vegetables from Florida to Eastern Points,* (219 I.C.C. 206, 209), wherein the Commission held:

> "Vegetable traffic from Florida to Eastern cities represents a substantial percentage of the total, and reduction to clear the fourth section would result in loss of greater revenue than would be gained from the additional traffic in the Wilmington district. Through rates from Florida to eastern destinations without observing aggregate-of-intermediates provision authorized."

Fancy Farm Case

An interesting and most unusual ruling in connection with aggregate of intermedate rates was made by the Commission in *Bacon Brothers, et al, v. Alabama Great Southern R. Co., et al,* commonly called the *Fancy Farm Case,* decided January 5, 1948.

The history of the case began with the publication of Agent Kipp's Tariff I.C.C. No. A-3267, effective July 1, 1941, naming joint commodity rates on fruits and vegetables from stations in Minnesota, North Dakota, South Dakota, Iowa, Nebraska, Wyoming, Colorado, Utah, Idaho and Oregon to stations in the Mississippi Valley and Southeastern territories. In this tariff, Fancy Farm, Ky., was erroneously assigned to Rate Group 1602 which was the group number assigned to six stations in northeastern Nebraska on the C.B. & Q. Railroad. The short-line distance between the Nebraska stations in Rate Group 1602 range from 5 to 23 miles, while the short line distance from Fancy Farm to the nearest point in Group 1602 is about 690 miles. The fact that the rates published in this tariff to Fancy Farm were unquestionably a tariff error was established beyond a reasonable doubt.

During the period that the erroneous commodity rates were in the tariff, normal commodity rates were also in effect on most of the traffic from the origin points to Fancy Farm and aggregate of intermediate rates composed of these normal rates to Fancy Farm with class rates beyond were higher than the joint through rates. However, the use of erroneous commodity rates to Fancy Farm and class rates beyond resulted in charges substantially under the joint through rates or aggregate of normal commodity rates to Fancy Farm and class rates beyond. The carriers assessed charges based on the joint through rates and the shippers requested reparation based on the charges which would have resulted from the use of the erroneous rates.

Division 2 of the Commission in the original report in these proceedings (263 I.C.C. 587, 590), stated that the defendants had taken the position that the publication of the rates to Fancy Farm should be found a void publication because the application of the combinations of rates sought would result in changing the relation of origin groups and so result in violations of the Commission's orders in *Leonard, Crosset & Riley, Inc. v. Akron C. & Y. Ry. Co.*, (176 I.C.C. 309), and other related proceedings, in which the Commission prescribed the rate relation of origins in Western Trunk Line territory with respect to carload rates on potatoes to points in the Southeast. Division 2 disagreed with defendants and said:

> "A shipper is not required to read the reports of this Commission before he can rely upon rates clearly published in the tariffs. If the carriers publish in the legal manner, rates which violate our orders, they may lay themselves open to prosecution therefor, but we have never found that such publication is void. The tariffs, established in lawful manner, and not our orders, state the rates that are legally applicable under Section 6 of the Act. The proposition that rates must be applied in accordance with the applicable tariffs, and that an error in tariff publication affords no legal ground for a departure from the tariff provisions, is so well settled as to require no citation of authorities."

In a later report (266 I.C.C. 303) on reconsideration by the entire Commission, they completely reversed the finding of Division 2, in the original report, and stated that inasmuch as the erroneous rates to Fancy Farm were not sought on traffic consigned to Fancy Farm but were used in constructing aggregate of intermediate rates under tariff rules, and that the erroneous rates were, in some instances, 50 percent of the normal rates, and were in violation of Sections 3 and 4 of the Act, they concluded that the correct rates were the normal rates for the purpose of applying the aggregate of intermediates rule and not the erroneous rates relied upon by the complainants.

Commissioners Alldredge and Aitchison dissented and in their report stated that the Commission's action was a reversal of the rule that had been followed for many years, that the shipper is entitled to the lower of two conflicting rates. Their report said that the decision of the majority invites the defense of tariff error in every case involving conflicting rates and thereby fosters litigation. They recommended that the prior findings of Division 2 be affirmed.

The Commission again reopened the proceedings for reconsideration, and in its report of January 5, 1948 (269 I.C.C. 571), stated that since the prior report situations had been disclosed where the erroneous rates were the only commodity rates to Fancy Farm, although class rates were published, and ordered that in those instances the erroneous commodity rates were applicable in constructing aggregate or combination rates and ordered reparation accordingly. The report affirmed the prior findings to the extent that where two commodity rates were published to Fancy Farm, one the correct rate and the other the erroneous rate, that the correct rate was the applicable rate in constructing aggregate of intermediate rates.

Commissioners Alldredge, Aitchison and Barnard joined in a dissenting report, and among other things, stated:

> "The rule that the lower of two conflicting rates is the legal charge is based upon the further principles that the intention of the framers of the tariff does not govern in the interpretation of that tariff and that tariff ambiguities must be resolved against the framers of the tariff."

Their dissenting report also pointed out that the title page of the tariff stated that it named rates to stations in Kentucky and there was nothing thereon to indicate any restrictions or limitations as to the Kentucky stations included.

To sum up, this report of the Commission with respect to the application of erroneous commodity rates in constructing aggregate or combination rates provides:

(a) Where such erroneous commodity rates are the only commodity rates applicable to the traffic to the point in question, such erroneous commodity rates are applicable even though class rates are in effect or joint through commodity rates are published to the ultimate destination.

(b) Where both an erroneous commodity rate and a proper or legal commodity rate is published, the proper or legal commodity rate is the applicable rate. This ruling applies only in constructing aggregate of intermediate rates and is not to be construed as applying where two commodity rates are published to the ultimate destination. In those instances the lower of the two rates is the applicable rate. For example, if Fancy Farm had been the ultimate destination of any of the shipments in the involved proceedings, the lower or erroneous rates would apply on these shipments regardless of whether there was also a correct commodity rate published.

It is difficult to understand how the Commission, in three attempts, could come up with so tenuous a result. In subsequent decisions involving tariff errors they have ruled that the tariff in effect when the shipment moved was binding upon the carrier and shipper alike, and an error in the publication of a rate therein affords no legal ground for a departure from the tariff provisions. Also, where there is a conflict or where a reasonable doubt exists as to the applicability of two disputed bases, a shipper is entitled to the benefit of the lower charge. The question of normal versus erroneous rates has no significance nor does the fact that the rate is used in a combination in any way alter the situation.

Power of the I.C.C. to Grant Relief

The original fourth section of the Act contained only the long-and-short-haul feature, and the rulings of the Commission prior to that time in declaring a through rate which exceeded the aggregate of the intermediates as being prima facie unreasonable was by virtue of the plenary powers in respect to rates vested in it under sections one, two, and three of the Act.

In interpreting this amendment, the Commission, in its *Annual Report to Congress* for the year 1911, stated:

> "By the act approved June 18, 1910, the fourth section ordinarily known as the 'long-and-short-haul clause' was amended. The amendments thereto have raised important questions and have cast upon the Commission a great amount of work, and it seems proper to indicate in a general way what is being done under the amended section.
>
> "The original section provided that no carrier should charge more for the short than for the long distance under 'substantially similar circumstances and conditions,' and one amendment of June 18, 1910, was the omission of these words. A new provision also was inserted prohibiting carriers from charging 'any greater compensation as a through rate than the aggregate of the intermediate rates subject to the provisions of this Act.'
>
> "Under the terms of the amended section the Commission is given authority to relieve carriers from the operation of the statute, and one of

the first questions presented was whether this power of relief extended both to the charging of the higher intermediate rate and to the charging more for the through service than the sum of the intermediate rates. Every practical reason requiring the exercise of the relieving authority in the first instance applies, apparently, in the second, and the Commission was of the opinion and held that it might grant permission to charge more for through service than the combination of the intermediate rates, as well as to impose a higher rate at the nearer point."

In analyzing and interpreting this provision of the fourth section, we come to a very fine point of law, involving the jurisdiction of the Commission, upon which many authorities in the traffic field held (prior to the *Patterson v. L. & N.* case which will be referred to later) that the fourth section of the Act as amended did not give the Commission authority to permit departures from the aggregate of intermediates provision. The specific question as to whether or not the Commission has such authority had not been before the United States Supreme Court, hence the court had not definitely passed upon this point. However, the Supreme Court, in the case of *J.W. Patterson, et al, v. L. & N. R.R. Co., et al.*, (269 U.S. 1), definitely held that the Commission has jurisdiction over the aggregate of intermediates provision of the fourth section insofar as the feature of the filing by the carrier of an application for relief is concerned, as follows:

"The shipper's contention that relief from the operation of the aggregate of intermediates clause was not invoked by an adequate and timely application is also unsound. Under the second proviso of section 4, the rates complained of, if in effect on June 18, 1910, and then unlawful, remained so, provided an application to suspend the operation of the section was duly made and was either allowed or remained undetermined. The District Court construed the report of the Commission as finding that the then existing rates here in question were so protected. We, also, construe the report as finding, in effect, that application for relief was made and was both adequate and timely."

Although the Supreme Court, in the same case *(Patterson v. L. & N.)*, did not make a positive statement, it made an extremely strong inference that the Commission also has authority to authorize departures where the through rate is higher than the aggregate of intermediates, in the following language:

"The aggregate of intermediates clause was inserted in section 4 by the Act of June 18, 1910. Since that amendment, as before, the section empowers the Commission, upon special application to prescribe the extent to which such designated common carrier may be relieved from the operation of this section. The question whether, after the amendment, the power so conferred was still limited to the long-and-short-haul clause or extended also to the aggregate of intermediates clause, received careful consideration immediately after the passage of the 1910 Act. The Commission concluded that its power to grant the relief applied to both these clauses. In its annual report for 1911 the reasons for this conclusion were set forth (pages 19-20). The construction then adopted has been acted upon consistently ever since. So far as appears, no court, federal or state, has taken a different view. And Congress has acquiesced."

In recent years the Commission has, in numerous cases, granted relief from the provisions of the aggregate of intermediates clause of the fourth section of the Act. For example, in *Iron and Steel from Official to W.T.L. Territory,* (318 I.C.C. 449), they found that although proposed rates on heavy iron and steel articles from Official Territory origins east of Illinois-Indiana border to W.T.L. destinations north and east of Missouri River are justified by compelling market competition with Chicago, some departures from the aggregate of intermediates clause will occur over routes through W.T.L. Territory to Official Territory points at or near the territorial border, where the single factor Official Territory rates exceeded combinations of the reduced rates to a W.T.L. point plus the W.T.L. rate beyond. As the competitive situation at the Official Territory destinations was sufficiently different to warrant continuance of the present higher rate level from the involved origins, grant of relief sought was justified.

As indicated previously, the present statute, P.L. 95-473, contains an aggregate of intermediates provision; however, there is nothing in Section 10726 authorizing the Commission to grant relief from the aggregate of intermediates clause contained therein.

Charging of Through Rate Higher Than Intermediates Not Overcharge

It should be thoroughly understood that the charging of a through rate which is higher than the aggregate of intermediate rates is not an overcharge. It represents a violation of Section 10726 of P.L. 95-473 and recovery of the difference between the higher through rate and the lower combination can only be accomplished through the medium of complaint to the I.C.C. or the filing of a Special Docket Application by the carrier requesting authority to award reparation.

Section 10761 of P.L. 95-473 prohibits the carrier from charging other than the rates and charges published in the carrier tariffs and on file with the Commission and Rule 55 of Tariff Circular 20 makes the through rate the only legal rate. Therefore, unless the carrier's tariff provides otherwise, the through rate must be applied whether or not it exceeds the aggregate of intermediates. Furthermore, since an overcharge is the assessment of a rate or charge higher than the applicable rate or charge published in the carrier's tariff, it follows that charging a published through rate can never be an overcharge even though it is higher than the aggregate of intermediates. It would violate the Act and therefore be unlawful, but it would be the applicable or legal rate.

In several early decisions the Commission and the courts ruled that a shipper or consignee could only obtain reparation for a violation of Section 4 upon a showing of actual pecuniary loss. As far as the long and short haul clause is concerned, this rule is applicable today. However, where the aggregate of intermediates clause is involved both the Commission and the courts have consistently held that through single factor rates which exceed the aggregate of intermediate rates are prima facie unreasonable. Strictly on the basis of this rebuttable presumption it would be necessary for a complainant to prove unreasonableness under Section 10701 of P.L. 95-473 in order to obtain reparation. However, as we have already seen, Rule 56(c) of Tariff Circular 20 specifically authorizes the payment of reparation in such cases.

In *C.A. Glass Co. v. Pacific Electric Ry.*, (265 I.C.C. 541), the Commission ruled:

> "Although the factors composing the combinations are depressed by truck competition, under the holding out of Rule 56 of Tariff Circular 20, shippers are entitled to reparation to the basis of the aggregate of intermediate rates, regardless of the level of those rates."

Also, in *Consumers Power Co. v. A. & S.*, (286 I.C.C. 291), the Commission said:

> "Provisions of Rule 56, where published in a tariff, apply to any situation where a rate published in that tariff is in excess of an aggregate of intermediate rates, however made, in violation of Section 4."

The inclusion in current tariffs of an aggregate of intermediate clause permitting the application of any combination of rates that defeats the through rate eliminates the necessity of paying the higher through rate and subsequently obtaining reparation.

Aggregate of Intermediate Rates Versus Joint Through Rates Via Motor Carriers

Part II of the Interstate Commerce Act, applying to motor carriers, did not contain an aggregate-of-intermediate rates provision as did Part

I, Section 4, applying to rail and water carriers, and Section 10726 of P.L. 95-473 which contains the present aggregate of intermediates provision, specifically restricts it to railroads and water carriers. Therefore, unless the motor carrier tariffs naming the joint through rates contain a provision requiring or permitting a lower aggregate of intermediates to apply, the long-established rule in the construction of rate tariffs, i.e., in the absence of appropriate tariff provisions to the contrary, a specific joint rate between two points is the only legal rate on the particular commodity over authorized routes between those points, even though some combination of intermediate rates may produce lower charges, must be applied. *Ingalls v. Maine Central R. Co.,* (24 Fed. (2d) 113; *Morgan v. Missouri, K & T. Ry. Co.,* (12 I.C.C. 535); *North American Cement Corp. v. Western Maryland Ry. Co.,* (129 I.C.C. 90).

Very few motor carrier tariffs contain an aggregate of intermediates provision permitting such lower basis to apply. However, Section 10726 of P.L. 95-473 provides:

> "A rate, classification, rule or practice related to transportation or service provided by a carrier subject to the jurisdiction of the Interstate Commerce Commission under Chapter 105 of this title must be reasonable."

Certainly under this provision of the Act, every joint through rate, assessed by common carriers by motor vehicle, that exceeds the aggregate of intermediate rates is prima facie unreasonable and unlawful and therefore prohibited. The Interstate Commerce Commission has so found in the case of *Kingan and Company v. Olson Transportation Company, et al.,* (32 M.C.C. 11, 12), wherein it held:

> "Part II does not contain an aggregate-of-intermediate-rates provision as does part I (Section 4). But prior to the enactment of that provision in 1910 (Mann-Elkins Act), the Commission held that the burden of proof was upon rail carriers to defend the reasonableness of a joint through rate that exceeded the aggregate of intermediate rates between the same points over the particular route. See *Patterson v. Louisville & N.R. Co.,* (269 U.S. 1). The same principles that governed rail rates under such circumstances prior to the 1910 amendment are applicable to similar conditions involving motor-carrier rates. Therefore, the assailed joint rates are presumed to have been unreasonable to the extent that they exceeded the corresponding aggregate of intermediate rates."

Also, in the case of *Victory Granite Co. v. Central Truck Lines, Inc.,* (44 M.C.C. 320, 324), the Commission held:

> "Joint rates which exceed the aggregate of intermediate rates are prima facie unreasonable. This presumption can be overcome only by a clear showing that the aggregate of intermediate rates is unreasonably low. The

defendants have failed to make such a showing here. They contend that it would be inconsistent to find unreasonable the joint second-class minimum rate, and at the same time to approve as a factor of the aggregate of intermediates, the fourth-class rate from Minneapolis to Lake City which itself results from a minimum rate restriction. The finding herein does not constitute an approval of the measure of the defendants' rate on polished building granite from Minneapolis to Lake City, or the manner of its publication. The conclusion reached is simply that the joint rate charged was unreasonable to the extent that it exceeded the combination of intermediate rates which would have applied on the shipment in the absence of a joint rate."

From the foregoing citations we observe that while the statute does not contain an aggregate-of-intermediate-rates provision applicable to motor carriers, the Commission has applied the prohibition of that provision to motor carriers on the basis of the broad requirement that rates and practices must be reasonable.

Steps Necessary to Break Down the Through Rate

It is sometimes advantageous to legally "break down the through rate." In proceeding to "break down the through rate" there are several features to be taken into consideration. In order to bring out these features in their logical sequence and in a manner most easily understood, it will be best to take a given situation as an example.

Let us assume, therefore, that there is a through local rate of 92 cents per 100 pounds in effect from Creston, Iowa to Chicago, Illinois. As Creston is a local point on the Burlington Northern, this rate is applicable only via that line. In checking this situation we might find a local or proportional rate published by the B.N. from Creston, Iowa to Burlington, Iowa of 41 cents (filed with the Commission and therefore applicable on interstate business) and a rate of 38 cents from Burlington, Iowa to Chicago, Illinois published by the C.R.I. & P., making a combination rate of 79 cents per 100 pounds.

In this case, in order to break down the through rate of 92 cents and secure the benefit of the lower combination of 79 cents, it would be necessary to bill the shipment *via the route over which the combination applied,* care being taken to avoid the route via which the published through rate was applicable. Therefore, by routing the shipment via the B.N. to Burlington, Iowa and C.R.I. & P. beyond to destination, instead of B.N. direct, the rate via the route over which the shipment moved would have to be protected and the lower combination would apply. In other words, as the C.R.I. & P. was not a party to the published through rate, such higher through rate would, because of its inapplicability to the route of actual movement, be broken down, thus leaving the combination rate as the only rate which could properly be used.

The important points in connection with the breaking down of through rates are:

(1) Is the combination of intermediates lower?

(2) Can the shipment be routed so that such lower combination will apply and the through rate will not?

If such a condition is found to exist, the shipment must be routed against the tariff naming the through route and through rate and in lieu thereof, sent via the route over which such lower combination applies. By routing the shipment against the tariff containing the published through rate in this manner, the applicability of that tariff is broken down, thus leaving the combination through rate as the only legal rate to be used.

In the event it is not possible to handle a violation of the aggregate of intermediates provision in the manner indicated, i.e., by sending the shipment over a route contrary to the tariff containing the published through rate and in accordance with a route via which the lower combination only is properly applicable, then the through rate must be attacked either by requesting the carriers to reduce the through rate to the sum of the intermediate rates or by complaint to the I.C.C. alleging violation of Sections 10726 and 10701.

In such cases the published through rate is, in view of the Commission's holdings, the only rate which may be legally assessed. However, as previously stated, since such higher through rate is unlawful (unless protected by an Order issued by the Commission), prompt relief therefrom may be obtained, through lawfully recognized channels, by requesting the carrier or carriers involved to reduce, on one day's notice, the through rate to equal the sum of the intermediates, as provided for in Rule 56 of the Commission's tariff circular.

If and when this is done, reparation, with interest, for the difference between the published through rate and the combination through rate on all shipments which had moved under the higher through rate during the two years preceding its reduction would be requested of the Commission. Under the recognized procedure in cases of this nature, the matter is submitted to the Commission, usually by the carrier at shipper's request, for authority to make reparation to complainant. As the facts surrounding these aggregate of intermediates violations are so obvious, and as the Commission has held such violations as prima facie unreasonable, these matters are usually handled by correspondence, on what is known as the "Informal Docket" or "Special Docket," without the necessity of a formal hearing, and upon a proper statement of facts, including proof of damage, reparation is usually awarded with little or no difficulty.

Via motor carriers we do not find the same conditions existing, because P.L. 95-473 does not contain an aggregate of intermediates provision applicable to motor carriers. However, we are still confronted with the necessity of knowing how to break down through rates.

Provision is made in the statute for the establishment of joint rates between motor carriers, also between motor carriers on the one hand and water lines or railroads on the other. In connection with such joint rates, the Commission has held that a through motor carrier rate, when once established, becomes the only legal rate for through transportation whether higher or lower than the aggregate of intermediate rates. Should a situation be encountered where the through rate exceeds the aggregate of intermediate rates, the only way the lower aggregate of intermediate rates can be used is by routing against the tariff naming the through rate.

If it is found to be impossible to route the shipment so that the lower aggregate of intermediate rates will apply, the higher through rate must be paid. In such a case the shippers only recourse would be to file a complaint with the Commission based upon unreasonableness if the carriers refused to reduce the through rate.

P.L. 95-473 permits the awarding of reparations for the exaction of unreasonable motor carrier rates. In the event that a combination of motor carrier rates is lower than the through rate, a shipper may sue in court to recover the difference. The suit is held in abeyance pending a determination by the I.C.C. as to whether the through rate is in fact unreasonably high, and the Commission's findings are used by the court in disposing of the case.

CHAPTER 8

THE LONG AND SHORT HAUL CLAUSE

General

Having examined the prohibition against charging more as a through rate than the aggregate of intermediate rates we shall now take a close look at the other prohibition of Section 10726, namely, the "long and short haul clause" which makes it unlawful for railroads and common water carriers, to "charge or receive any greater compensation for the transportation of property of the same kind or of passengers for a shorter distance than for a longer distance over the same line or route in the same direction, (the shorter distance being included in the longer distance)."

As an aid to proper interpretation of this provision, let us consider a situation involving a possible violation of the long and short haul clause as indicated by the following diagram:

A ———————— B ———————— C ———————— D

Here we have a common carrier hauling goods in interstate commerce between points "A" and "D," and passing through "B" and "C."

As stated above, the long and short haul provision specifies that it shall be unlawful for such common carrier to charge more for transporting like kind of freight (over the same route in the same direction) from "A" to "C" than from "A" to "D." It may, if circumstances are such as to warrant it, charge the same rate from "A" to "C" as it charges from "A" to "D," but it must not charge more. Likewise, it would be unlawful for the carrier to charge less for hauling a shipment of the same kind from "A" to "C" than from "B" to "C."

Each of the situations described in the preceding paragraph would, under normal circumstances, constitute a violation of Section 10726. There is, however, another provision in Section 10726 that authorizes the Commission, after investigaion, to permit a carrier to charge less for longer distances than for shorter distances and where this occurs there would be no violation but rather an authorized departure from the provisions of Section 10726.

Prior to October 10, 1930, the Commission, under authority of Tariff

Circular Rule 77, authorized carriers to publish commodity rates from known points of production (or to known points of consumption) without making such rates applicable from (or to) unnamed intermediate points. Under such circumstances, when a shipment moved from (or to) an intermediate point (under the higher class rate) higher charges were assessed from (or to) such intermediate point than from (or to) the point beyond. Rule 77 was a promise on the part of the carriers to publish a commodity rate on one day's notice, from (or to) such intermediate point that would not exceed the commodity rate applicable from (or to) the next more distant point. Rule 77 also required the carriers to request the Commission to authorize a refund of the difference between the rate and charges assessed and those subsequently established. The primary object of Rule 77 was to prevent violations of the long and short haul clause. Since the cancellation of this Rule became effective, the railroads have been providing intermediate application in their tariffs, as authorized in Tariff Circular Rule 27, as a means of preventing long and short haul violations.

In many instances carriers charge the shipper more for the shorter haul than they do for the longer haul—a situation which would, without extenuating circumstances, be unreasonable and therefore in violation of the law. In the following sections we will discuss this matter and see that circumstances and conditions will justify the charging by the carrier of more for the smaller service of transportation miles which it sells the shipper than it charges for the greater service, as well as the means which the law has provided for a practical determination of those instances in which the carrier may continue practices of this nature, the successive growth and amendment of the long and short haul clause to its present form, and the restrictions which have been placed upon the Commission by Congress in authorizing it to permit departures from this restriction of the statute.

Early History of the Clause

In order to get a clear conception of that part of the fourth section of the Interstate Commerce Act known as the "long and short haul clause," we must first go back to the Act as it was originally enacted in 1887. When the *Act to Regulate Commerce* was first put on the statute books, the fourth section read as follows:

> "That it shall be unlawful for any common carrier subject to the provisions of this Act to charge or receive any greater compensation in the aggregate for the transportation of passengers or of like kind of property,

> *under substantially similar circumstances and conditions,* for a shorter than for a longer distance over the same line, in the same direction, the shorter being included within the longer distance; but this shall not be construed as authorizing any common carrier within the terms of this Act to charge and receive as great compensation for a shorter as for a longer distance. *Provided,* however, That upon application to the Commission appointed under the provisions of this Act, such common carrier may, in special cases, after investigation by the Commission, be authorized to charge less for longer than for shorter distances for the transportation of passengers or property; and the Commission may from time to time prescribe the extent to which such designated common carrier may be relieved from the operation of this section of the Act."

Congress realized, however, that an absolute rule of this nature could not be enforced without causing injury to some industries and carriers. It therefore specified in this section that the movements would have to be "under substantially similar circumstances and conditions." It also granted the Commission authority, in special cases, upon investigation, to authorize departure from this restriction of the law.

In commenting upon the reasons underlying enactment of the long and short haul clause, and the necessity for departures therefrom, the Commission, in its First Annual Report to Congress, stated:

> "The provisions against charging more for the shorter than for the longer haul under the like circumstances and conditions over the same line and in the same direction, the shorter being included within the longer distance, is one of obvious justice and propriety. Indeed, unless one is familiar with the conditions of railroad traffic in sections of the country where the enactment of this provision is found to have its principal importance, he might not readily understand how it could be claimed that circumstances and conditions could be such as to justify the making of any exceptions to the general rule.
>
> "It is a part of the history of the Act that one house of Congress was disposed to make the rule of the fourth section imperative and absolute, and it is likely that in some sections of the country many railroad managers would very willingly have conformed to it, because for the most part they could have done so without loss, and with very little disturbance to general business. But in some other parts of the country the immediate enforcement of an ironclad rule would have worked changes so radical that many localities in their general interests, many great industries, as well as many railroads, would have found it impossible to conform without suffering very serious injury. In some cases probably the injury would have been irremediable. To enforce it strictly would have been, in some of its consequences in particular cases, almost like establishing, as to vested interests, a new rule of property."

The Commission felt at that time that a study of the conditions under which railroad traffic in certain sections of the country had sprung up was necessary to an understanding of the difficulties which surrounded the subject. The southern territory bounded by the Ohio and Potomac rivers on the north and by the Mississippi river on the west presented to the Commission an opportunity and also an occasion, for such study. The railroad business of that territory had grown to what it was in sharp competition with water carriers, who not only had the ocean at their service, but by means of navigable streams were able to penetrate the interior in all directions. The carriers by water were first in the southern territory and were doing a very thriving business while railroads were coming into existence; but when the railroads were built the competition between them and the water carriers soon became sharp and close, and at the principal competing points the question speedily came to be, not what the transportation service was worth, or even what it would cost the agency performing it, but at what charge for its service the one carrier or the other might obtain the business.

In this competition the water carriers had great advantages; the capital invested in their business was much smaller; they were not restricted closely to one route, but could change from one to another as the exigencies of business might require; and the cost of operation was less. However, the railroads had an advantage in greater speed, which at some times, and in respect to some freight, was controlling.

Because of this competition of water carriers and rail carriers the transportation rates which were directly controlled by the water carriers soon reached a point to which the railroads could not possibly have reduced all their rates and still maintained a profitable existence. The rail carriers did not attempt such a reduction, but on the contrary, while reducing their rates at the points of water competition to a level that should be necessary to enable them to obtain the traffic, they kept the rates up at all other points to such a level as they deemed the service to be worth, or as they could obtain. It often happened, therefore, that the rates for transporting property over the entire length of a railroad to a terminus served by a water carrier would not exceed those for the transportation for half the distance only, to an intermediate or way station not similarly favored with competition.

The seeming injustice of this situation was excused by the railroads on the plea of necessity. They argued that the rates to the terminus were fixed by the competition and could not be advanced without abandonng the business to the water carriers. Further, that the higher rates to the local intermediate points were no more than was reasonable, and they were not, by reason of the low rates to the competitive point, made higher than they otherwise would have been. Also, if the rates on the railroad were established on a mileage basis throughout, with no regard to special competitive forces at particular points, the effect in diminishing the volume of business would be so serious that local rates at non-competitive points would necessarily be advanced beyond what they were made when the competitive business was taken at rates which left little margin above the actual cost of movement. Such was the common argument advanced by all railroads in support of the higher rates to short haul points.

The foregoing situation was not due altogether to water competition; railroad competition had been allowed to have a similar effect in establishing lower rates on the longer haul. Also, to some extent, lower rates had been conceded to important cities in order to equalize advantages as between them and other cities which were either rivals or to which lower rates had been given under a pressure of necessity. Lower rates were given also in many instances as a means of building up a long-haul traffic that could not possibly bear the local rates, and which consequently would not exist at all if rates were established on a mileage basis, or on any basis which, as between the long and short haul traffic, undertook to preserve anything like the relative equality.

In commenting upon the effect of the foregoing situation, the Commission, in its First Annual Report, stated:

> "It would be foreign to the purposes of this report to discuss at this time the question whether in this system of rate-making the evils or the advantages were most numerous and important. Some of the evils are obvious; not the least of which is the impossibility of making it apparent by those who have not considered the subject in all its bearings, that the greater charge for the shorter haul can in any case be just. The first impression necessarily is that it must be extortionate; and until that is removed it stands as an impeachment of the fairness and relative equity of railroad rates. But, on the other hand, it must be conceded that this method of making rates represents the best judgment of experts who have spent many years in solving the problems of railroad transportation; and its sudden termination without allowing opportunity for business to adapt itself to the change would, to some extent, check the prosperity of many important places, render unprofitable many thriving enterprises, and probably put an end to some long-haul traffic now usefully carried on between distant parts of the country. It is also quite clear that the more powerful corporation of the country, controlling the largest traffic and operating on the chief lines

> of trade through the most thickly settled districts, can conform to the statutory rule with much more ease and much less apparent danger of loss of income than can the weaker lines, whose business is comparatively light and perhaps admits of no dividends, and the pressure of whose fixed charges imposes a constant struggle to avoid bankruptcy."

It will be observed from the reasons given in the above citation that the underlying purpose of the long and short haul clause of the fourth section was to to end what was felt to be arbitrary discrimination in transportation charges as between cities, resulting from the efforts of railroads to meet competition. By plainly prohibiting a practice of this nature, the statute prevents a large number of abuses which would exist with impunity until separately condemned in actions brought under the first or third sections of the Act. It is analogous, in effect, to the old but always applicable adage that "an ounce of prevention is worth a pound of cure."

Early Application of Long and Short Haul Clause

It will be seen from the original fourth section, as quoted earlier in this chapter, at that time the prohibition against exacting a greater charge for the shorter than for the longer distance over the same line in the same direction was modified by the words "under substantially similar circumstances and conditions." Immediately after the adoption of the Act, the Commission, in considering the meaning of the law and the scope of the duties imposed on it in enforcing the fourth section, reached the conclusion that the words "under substantially similar circumstances and conditions" dominated the long and short haul clause and empowered carriers to primarily determine the existence of the required dissimilarity of circumstances and conditions, and consequently to exact in the event of such difference a lesser charge for the longer haul than was exacted for the shorter haul. It also felt that competition which materially affected the rate to a particular point was a dissimilar circumstance and condition within the meaning of the Act. This left the carrier at liberty to act on its own judgment as to the conditions and circumstances in any case, subject to responsibility to the law in the event a greater charge were made for the shorter distance when the circumstances and conditions were in fact not dissimilar.

It was not until the year 1892 that the Commission decided cases upon the principle that the carriers should not be the judge and jury in determining whether or not the circumstances and conditions were dissimilar, but should apply to the Commission, who, after due investigation, would decide the matter. These rulings of the Commission were contested by the carriers in the courts, and, in 1897, the Supreme Court of the United States, in the case of the *Interstate Commerce Commission v. Alabama Midland Railway Company, et al.*, (168 U.S. 144), held that competition between carriers was an effective circumstance that brought about substantially dissimilar circumstances and conditions, and that such competition, *when controlling,* would justify the charging of the lower rate to the farther distant point, *not as a matter of grace or favor of the Commission but as a matter of right.*

This construction of the Act left it entirely to the railroads to determine whether the competition which existed created the dissimilar circumstances and conditions provided for by the fourth section, and they were, then, of course, under such determination, permitted to charge more for the shorter than for the longer haul.

In referring to this interpretation of the Act by the Supreme Court, the Commission, in its Annual Report to Congress for the year 1897, stated:

> "That section (the fourth) exacts that the carrier shall not charge more for the short haul than for the long haul under substantially similar circumstances and conditions. If the circumstances and conditions are similar, the greater charge can not be made. If the circumstances and conditions are not similar, the section does not apply. The court holds that railway competition of controlling force makes the circumstances dissimilar. If, therefore, we find in a particular case that competition of controlling force actually exists, that ends the matter. We have no power to say whether, nor to what extent, such competition justifies the higher rate to the intermediate point."

Under this holding of the Supreme Court the long and short haul clause became inoperative insofar as restraint by the Commission was concerned. The carriers hereafter decided in every instance whether they could violate this provision of the law by ascertaining for themselves whether the conditions and circumstances were similar or not, and where it was to their advantage to do so, you may rest assured that they found the circumstances and conditions dissimilar.

Restoration of Commission's Jurisdiction in 1910

The amendment of June 18, 1910, had the effect of restoring the jurisdiction of the Commission by elimination of the words "under substantially similar circumstances and conditions." That is to say, a carrier should not thereafter, under the amended law, judge for itself what constituted dissimilar circumstances and conditions but was required to apply to the Commission for a determination by that body of whether the lower charge to the more distant point was justified by the circumstances of the case.

Full and unlimited powers to determine whether or not the long and short haul clause could be violated having been exclusively vested in the Commission in 1910, the law made it clear that *any violation not approved by the Commission was unlawful.* Since, however, thousands of fourth section violations had been made under the prior arbitrary actions of the carriers (which arbitrary actions, as heretofore indicated, were sanctioned by the various courts and particularly the Supreme Court), it was found necessary to provide in the amended section that any of then existing fourth section violations would have the effect of having been sanctioned by the Interstate Commerce Commission *provided* each carrier violating the fourth section would file with the Commission, prior to February 17, 1911, an application enumerating the particular violations, which application had the effect of tentatively authorizing the violation, even without investigation by the Commission, until that body could find time to investigate each case specifically.

In other words, in order for a fourth section violation to be lawful today, it must have been specifically investigated and authorized by the Commission.

The Commission has jurisdiction over interstate traffic only, therefore, it must be understood that any fourth section violations on intrastate traffic may or may not be permissible under the various state laws, although as a general understanding we may accept the fact that practically every state has a similar law.

In connection with this feature, the Commission in the *Meridian Rate Case,* (66 I.C.C. 179, 183), held:

> "These instances they (the complainants) characterize as fourth section violations, but inasmuch as they exist only in connection with intrastate traffic they would not constitute violations of the interstate commerce act."

There are many advocates who are in favor of a so-called rigid long-and-short-haul clause, to the effect that regardless of circumstances and conditions no carrier should be permitted to charge more for the shorter haul than for the longer haul. As developed from the citation taken from the Commission's First Annual Report to Congress, this is not a new thought but is one which was carefully considered at the time the Act was originally passed in 1887. The influencing considerations which existed then exist today; in fact, are even more potent in view of the increased commercial activity which characterizes present day business. Competition in our commercial life as well as that between carriers, not alone via all-rail but also between rail and water lines, has been a powerful influence in the tremendous growth of our country. In addition to the foregoing, the railroads now have to meet the competition of motor carriers. To assume that under no circumstances should the long and short haul clause be departed from would serve to disrupt the commercial and transportation growth which has been attained under the present state of recognized departures.

Limitations of Commission's Authority

When the Act to Regulate Commerce was amended in 1920 by the Transportation Act, the power of the Commission to permit departures from the long and short haul clause was made subject to three important limitations, as follows:

(1) That the Commission may not authorize the establishment of any rate or charge to or from a more distant point that is not *reasonably compensatory* for the service performed.

(2) That if a circuitous rail line or route is, because of such circuity, granted authority to meet the charges of a more direct line or route to or from the competitive points and to maintain higher charges, to or from intermediate points on its line, the authority shall not include intermediate points as to which the haul of the petitioning line or route is no longer than that of the direct line or route between the competitive points.

(3) That the Commission may not grant authority to violate the long and short haul clause on account of merely potential water competition not actually in existence.

In analyzing the first of these limitations, which is to the effect that the rate to the farther distant point must be reasonably compensatory, the thought naturally occurs that if the rate to such farther distant point is reasonably compensatory, then all the higher rates from or to intermediate points would be unreasonably high, and consequently unlawful in violation of Section 1 of the Act.

There is no definition in the Act of the term "reasonably compensatory" as contained in this limitation. In interpreting this term, the Commission, in the *Transcontinental Cases of 1922*, (74 I.C.C. 48, 71), stated:

"*** We are of the opinion and find that in the administration of the fourth section the words 'reasonably compensatory' imply that a rate properly so described must (1) cover and more than cover the extra or additional expenses incurred in handling the traffic to which it applies; (2) be no lower than necessary to meet existing competition; (3) not to be so low as to threaten the extinction of legitimate competition by water carriers; and (4) not impose an undue burden on other traffic or jeopardize the appropriate return on the value of carrier property generally, as contemplated in section 15a of the Act. It may be added that rates of this character ought, wherever possible, to bear some relation to the value of the commodity carried and the value of the service rendered in connection therewith. We also find that where carriers apply for relief from the long-and-short haul clause of the fourth section and propose the application of rates which they designate as 'reasonably compensatory,' they should affirmatively show that the rates proposed conform to the criteria indicated above. It goes without saying that carriers should not propose rates or rate structures for approval in a fourth-section application which create infractions of other provisions of the Interstate Commerce Act, and particularly of section 3."

In *Meat Packing House Products to the South,* (313 I.C.C. 464), the Commission found the proposed rates reasonably compensatory when they in all instances exceeded out-of-pocket costs and in most instances exceeded fully distributed costs.

The second limitation, commonly known as the "equi-distant clause" was taken out of Section 4 in the amendment of September 18, 1940.

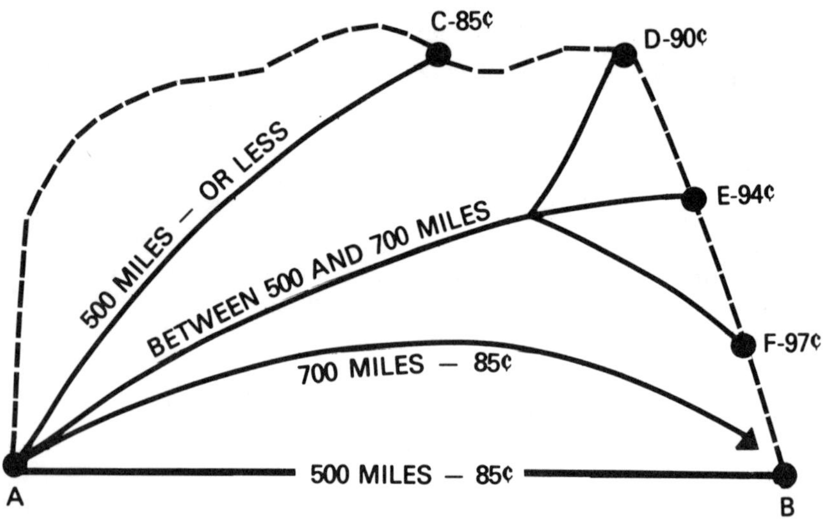

Although not now a part of the Act the equi-distant principle is frequently applied by the Commission when they believe it necessary to effect a reasonable adjustment of rates at intermediate points. The diagram on the previous page should be quite helpful in enabling you to get a clearer understanding of this feature:

Here we have a direct line distance from "A" to "B" of 500 miles, as indicated by the heavy line, for which the carrier charges a rate of 85 cents. The circuitous road, indicated by the dotted line, with a haul of 700 miles, desires to compete for business between these points at the same rate, without reducing its intermediate rates. In the event the Commission granted the circuitous line authority to meet the 85 cent rate at "B" the circuitous line, under the equi-distant principle, could not charge more than 85 cents to points on its line which are less distant from "A" than 500 miles; in other words, from "A" up to point "C" the circuitous line could not be permitted to violate the fourth section. The Commission might, however, grant authority to charge more at the intermediate points "D," "E," and "F," *which are farther distant from "A" than 500 miles.*

This principle appears to have been based on the theory that if the direct line can haul traffic from "A" to "B," a distance of 500 miles at a profit then there existed no good reason why the indirect road, if it wished to compete for this business, could not also haul intermediate traffic up to 500 miles on its line for at least the same rate.

In connection with this, the Commission held, in *Cohen-Schwarz Rail & Steel Co. v. M.L. & T. R.R.S.S. Co., et al.,* (59 I.C.C. 202, 203), *Coffee from Galveston, Texas., and Other Gulf Ports,* (58 I.C.C.716); *West Virginia Rail Co. v. I.C. Ry. Co.,* (53 I.C.C. 21, 27), and in other cases in volving the same principle, that when a request for relief was made based upon circuity, the longer route would be considered as being circuitous only when it exceeded the mileage via the direct route by fifteen per cent or more.

The Commission's interpretation of the third limitation is that if the necessary facilities are available to handle the traffic via water and a bona fide offer to transport has been made by the water carriers, the competition is actual and not potential even though there has been no movement of the commodity by water. For instance, in *Salt Cake to Southern Territory,* (237 I.C.C. 227, 230, 231), Division 2, of the Commission held:

> "The record establishes that all of the origins and destinations are located on water of sufficient depth to afford transportation by water, that

although the mills are of fairly recent construction, some shipments have moved by water, and that rates have been quoted by reliable carriers offering to transport the traffic between all of the points at rates, including the cost of additional handling and insurance, which are substantially lower than those proposed. In the circumstances, actual water competition exists. See *Rags and Paper to Newark, N.Y.*, (208 I.C.C.327)."

In the case cited in the Salt Cake decision quoted above, the Commission stated:

"In determining whether the competition by water is actual or is merely potential, 'not actually in existence,' the fact there has been no movement by water is significant but not controlling. In *Transcontinental Rates,* (46 I.C.C. 236), it was said that competition involves a striving between or among two or more persons or organization for the same subject. Competition in a commercial sense is the independent endeavor of two or more persons to obtain the business patronage of a third by offering more advantageous terms as an inducement to secure trade. The pertinent facts bearing upon the question of competition clearly bring the action of the interested carriers in the instant case, one of which has a water route, within this definition. The essential elements of competition are all present when a going water carrier has made a *bona fide* offer to perform a competitive service and is ready, willing, and able to carry the traffic if the offer is accepted, especially if every facility for the performance of that service is at hand. Nothing in that statute indicates that an actual movement of the particular commodity by water is necessary to establish the existence of water competition. Such movement or absence of movement only shows whether or not the water carrier was successful in obtaining traffic for which it was competing. As stated in *Pacific Coast Fourth Section Applications,* (129 I.C.C. 3, 15), the rail carriers actual water competition is the cost to the shipper of the water rate plus such incidental expenses as are met in using the water service. Here the water carrier has clearly shown the cost to the shipper if its offer to perform that service is accepted. We are of the opinion that actual water competition is present in the instant case."

Amendments Created by the Transportation Act of 1940

The Fourth Section was again amended by the Transportation Act of 1940, effective September 18, 1940, as follows:

"(Par. 1) It shall be unlawful for any common carrier subject to this part or part III to charge or receive any greater compensation in the aggregate for the transportation of passengers, or of like kind of property, for a shorter than for a longer distance over the same line or route in the same direction, the shorter being included within the longer distance, or to charge any greater compensation as a through rate than the aggregate of the intermediate rates subject to the provisions of this part or part III, but this shall not be construed as authorizing any common carrier within the terms of this part or part III, to charge or receive as great compensation for a shorter as for a longer distance: Provided, That upon application to the Commission such common carrier may in special cases, after investigation, be authorized by the Commission to charge less for longer than for shorter distances for the transportation of passengers or property, and the Commission may from time to time prescribe the extent to which such designated common carrier may be relieved from the operation of this section, but in exercising the authority conferred upon it in this proviso the Commission shall not permit the establishment of any charge to or from the more distant point that it is not reasonably compensatory for the service performed; and no such authorization shall be granted on account of merely potential water competition not actually in existence: And provided further, that tariffs proposing rates subject to the provisions of this paragraph may be filed when application is made to the Commission under the provisions hereof, and in the event such application is approved, the Commission shall permit such tariffs to become effective upon one day's notice.

"(Par. 2) Whenever a carrier by railroad in competition with a water route or routes reduce the rates on the carriage of any species of freight to or from competitive points it shall not be permitted to increase such rates unless after hearing by the Commission it shall be found that such proposed increase rests upon changed conditions other than the elimination of water competition."

It will be noted that *water carriers* were added to the provisions of the Section and that the power of the Commission to *permit* departures therefrom was changed by the elimination of the so-called equidistant clause. By the change it seems apparent that carriers may charge more via a circuitous route even though the distance to an intermediate point via such route may be less than to the competitive point via the direct line. In other words, the burden is always upon the carrier to justify the maintenance of a higher rate to *any* intermediate point than to the more distant point or points in any given case.

Changes Created by the Amendment of July 11, 1957

The Interstate Commerce Commission recommended to the Congress a Section Four bill (S.937) which would eliminate the requirement that the Commission approve in advance the publication of circuitous-route rates competitive with direct-route rates between the same points. The Congress made no change in the bill, reporting the proposal would "enhance the Commission's administrative effectiveness and would enable the carriers to simplify and improve their tariffs to the benefit of the carriers that publish and file the tariffs, the shippers that use them, and the Commission that must administer the terms of the tariff." The President signed this bill into law July 11, 1957.

Section 4(1), as amended by the *Amendment of July 11, 1957,* and the *Amendment of September 27, 1962,* reads as follows:

"(1) It shall be unlawful for any common carrier subject to this part or part III to charge or receive any greater compensation in the aggregate for the transportation of passengers, or of like kind of property, for a shorter than for a longer distance over the same line or route in the same direction, the shorter being included within the longer distance, or to charge any greater compensation as a through rate than the aggregate of the intermediate rates subject to the provisions of this part or part III, but this shall not be construed as authorizing any common carrier within the terms of this part or part III to charge or receive as great compensation for a shorter as for a longer distance: *Provided,* That upon application to the Commission and after investigation, such carrier, in special cases, may be authorized by the Commission to charge less for longer than for shorter distances for the transportation of passengers or property, and the Commission may from time to time prescribe the extent to which such designated carriers may be relieved from the operation of the foregoing provisions of this section, but in exercising the authority conferred upon it in this proviso, the Commission shall not permit the establishment of any charge to or from the more distant point that is not reasonably compensatory for the service performed; and no such authorization shall be granted on account of merely potential water competition not actually in existence: *Provided further,* That any such carrier or carriers operating over a circuitous line or route may, subject only to the standards of lawfulness set forth in other provisions of this part or part III and without further authorization, meet the charges of such carrier or carriers of the same type operating over a more direct line or route, to or from the competitive points, provided that rates so established over circuitous routes shall not be evidence on the issue of the compensatory character of rates involved in other proceedings: *And provided further,* That the tariffs proposing rates subject to the provisions of this paragraph requiring Commission authorization may be filed when application is made to the Commission under the provisions hereof, and in the event such application is approved, the Commission shall permit such tariffs to become effective upon one day's notice: *And provided further,* That the provisions of this paragraph shall not apply to express companies subject to the provisions of this part, and that the exemptions herein accorded express companies shall not be construed to relieve them from the operation of any other provision contained in this Act."

The principal change made by the *Amendment of July 11, 1957,* is embodied in the second proviso. By this change circuitous lines are permitted to meet the charges of the direct line or route without further authorization from the Commission, subject, however, to the standards of lawfulness set forth in other provisions of the Act and that rates so established shall not be evidence on the issue of the compensatory character of rates involved in other proceedings. The *Amendment of September 27, 1962,* exempts express companies from Section 4.

Long and Short Haul Violations of Motor Carriers

The long and short haul provision in Section 10726 of P.L. 95-473 applies only to carriers subject to the jurisdiction of the Interstate Commerce Commisison under subchapters I and III of Chapter 105, namely railroads and water common carriers. There is nothing in the present statute nor was there anything in the former Interstate Commerce Act making such a provision applicable to motor common carriers. However, the Commisison has held in numerous cases that such a situation is prima facie unreasonable and therefore violates the requirement of the statute that rates and practices must be reasonable. In *Fifth Class Rates Between Boston, Mass., and Providence, R.I.,* (2 M.C.C. 530), the Commission said:

> "Respondent states that it does not have a large volume of traffic to these intermediate points, gave no consideration to them when it published the proposed rates, and intends to remedy the situation later. The fact that traffic to such points is less than to the more distant points is obviously no satisfactory reason for the maintenance of higher rates to such points. In the absence of a sound reason, therefore, the maintenance of class rates to intermediate points higher than to more distant points on the same route by a motor carrier is prima facie unreasonable."

This position of the Commission with respect to long and short haul violations of motor carriers was affirmed in the case of *Hausman Steel Co. v. Seaboard Freight Lines, Inc., et al.,* (32 M.C.C. 31, 39).

In *Lynchburg Traffic Bureau v. Smith's Transfer Corp.,* (310 I.C.C. 503), the Commission held that:

> "Class rate on mixed LTL shipments of general merchandise from White Plains, N.Y. to Lynchburg, Va., higher than to more distant Virginia points over same route in same direction is prima facie unreasonable."

In a subsequent decision, *Class Rates, Chicago, Ill., to Texas,* (311 I.C.C. 660), they said:

> "Prima facie presumption that motor rates to intermediate points which exceed the corresponding rates to more distant points over the same route are unreasonable may be rebutted by a showing that the rate differences are warranted by differences in competitive conditions."

In connection with joint motor-rail rates the Commission in *Motor-Rail-Motor Traffic in East and Midwest,* (219 I.C.C. 245), said:

> "While Section 4 does not apply to the charges of motor common carriers, when such carrier joins in a through route and joint rates with a railroad it becomes a participant with the railroad in a movement which is subject to that section. Motor common carriers are not required to join with rail carriers in such routes and rates but, having voluntarily entered into such a joint arrangement, the motor carrier assumes obligations similar to those of the participating rail carrier in the observance of the provisions of Section 4."

In *Oil Country Iron or Steel Pipe, Midwest to Oklahoma and Texas,* (326 I.C.C. 511), the I.C.C., with several commissioners dissenting, ruled that joint motor rail rates are subject to the long and short haul provisions of Section 4 of the Act. This decision was upheld by the U.S. District Court in *A.T. & S.F. v. U.S. & I.C.C.,* (300 F. Supp. 1351), (1969).

When a motor carrier enters into an arrangement with a rail carrier for through carriage and relief is sought from the provisions of the fourth section, the Commission would be subject to the same limitations previously mentioned in granting such relief.

CHAPTER 9

TRANSIT PRIVILEGES

General

Transit privileges are services allowed by the carrier as the result of which a shipper may stop a shipment at some point intermediate between the origin and the destination of a shipment for a particular purpose. Transit privileges are provided on a wide variety of commodities and permit many different operations at the transit stations. A shipment may be stopped for storage, sorting, packing, mixing, milling, fabricating, manufacturing, blending and many other services.

The United States Supreme Court in *Central Railroad of New Jersey v. United States,* (257 U.S. 247), stated:

> "Transit rests upon the fiction that the incoming transportation to the transit point and the outgoing transportation from the transit point, which are in fact separate, distinct shipments, constitute a single, continuous shipment of the identical article shipped from the transit point, from origin to final destination."

The underlying purpose behind the establishment of the many various types of transit privilege is to permit a greater flexibility in the through movement of commodities requiring some processing between the time they leave the point of production and the time they are offered to the consumer. These privileges are not established at all points and under all circumstances and it is only where some good reason exists for the establishment of such privilege that it is permitted and the necessary rule published in the carrier's tariff.

In I & S Docket 5470, Transit and Mixing Rules on Foodstuffs, (270 I.C.C. 164), the Commission had the following to say regarding transit arrangements generally:

> "When the Interstate Commerce Act became effective, the maintenance of transit arrangements was a well established practice. The Commission was faced with the fact that to force the abolishment of such arrangements would destroy the value of enormous investments made on the understanding that they would be continued, and disrupt the marketing practices of some commodities, principally grain. The Commission, although somewhat hesitantly, found that within the proper limits transit arrangements were not violative of the Act.
>
> "Since that time numerous transit arrangements on the basis of the par-

ticular facts and circumstances attending each of them have been recognized by us as legal. Manifestly, however, there must be some limit to the application of this idea. That artifice is a departure from generally accepted and legitimate methods of rate making and should be sanctioned only when there are compelling reasons for its indulgence. It was adopted by us only as a matter of expediency and not as a device to enable carriers to avoid the requirements and restraints of the Act or rules and regulations published in tariffs governing the application of rates."

Transit is a service that generally may be granted or withheld by the carrier in its discretion so long as no unlawful discrimination results therefrom. However, the Commission has the power to require the establishment of transit. It is a privilege that must be specifically authorized in the carrier's tariff and all conditions and limitations therein prescribed with reference to it must be strictly observed. Section 10762 provides for the publication of transit privileges in carrier tariffs.

Furthermore, the Commission has held that the granting or withholding of transit is a matter local to the railroad on whose line the transit point is located.

Development and Abuses of Milling in Transit Privileges

The following paragraphs deal specifically with the milling-in-transit of grain since this is the oldest, and perhaps the most widely used transit privilege. With regard to the transiting of some other commodities specific rules could be encountered that would differ slightly from those applicable to the milling of grain, however it is felt that this discussion of milling-in-transit should provide the reader with most of the basic principles of transit.

As the advantages of these arrangements became apparent, the practice quickly spread to other points and when the Act to Regulate Commerce became a law in 1887 they had risen to such an exalted position in the economic development of the country that the Commission hesitated to condemn them, even though it was originally inclined to the belief that an arrangement of this kind which apparently impaired the integrity of the through rate, was unlawful. It well realized the tremendous hold that these privileges had secured on the commerce of the nation and that to abrogate them would lead to the virtual confiscation of millions of dollars in value which had been invested in property on the faith of their continuance. It therefore justified the preservation of the through rate in arrangements of this kind, despite the apparent reshipment involved, by holding that the movement was continuous and that when the article was being treated at the transit point, the service of transportation was temporarily "suspended," the product, when it finally went forward to point of consumption, being considered as completing the journey upon which it had entered when the raw material was taken up and the original contract of carriage entered into. This theory of "suspended" transportation was later upheld by the Supreme Court of the United States and the lawfulness of transit arrangements thus established.

As these privileges grew it was inevitable that under the stress of competition and commercial rivalry, and because of the many opportunities which transit afforded for doing so, abuses and discrimination should creep in. In the early part of 1908 these abuses were very forcibly brought to the attention of the Commission by numerous complaints from shippers to the effect that competitors at transit points were, by various substitutions of commodities and of billing, avoiding payment of the lawfully published rates.

In an attempt to put a stop to these practices, and in order to enforce the law as it then interpreted it, the Commission, on June 25, 1908, issued a ruling to the effect that it is unlawful to substitute at the transit point, or forward under the transit rate, tonnage or commodity that does not move into that point on that same rate.

Practically no attention was paid to this ruling by either shippers or carriers. The complaints continued to grow, however, and the abuses increased rather than decreased in volume. Therefore on June 29, 1909, the Commission promulgated Rule 76 of Tariff Circular 17-A, which stated it was not permissible at transit point to forward on transit rate a commodity that did not move into transit point on transit rate, or to substitute a commodity originating in one territory for the same or like commodity moving into the transit point from another territory, or to make any substitution that would impair the integrity of the through rate. The rule further provided that carriers would be expected to conform their transit rules and their billing to the suggestions of this rule. Failure of carriers to do so would be regarded as voluntary concessions from legal rates.

Following the establishment of this ruling, many complaints were made to the Commission by shippers at transit points. It was felt that Rule 76 was too strict and that its application would destroy all benefits of the transit privileges. An appeal was made to the Commission to investigate anew the question of substitution of tonnage in transit. For four years the Commission held extensive hearings, conferences, and investigations and after exhaustive study it reached the decision that the

Act to Regular Commerce did not, within certain limits, prohibit the substitution of tonnage and commodities at transit points, so long as substitutions were not unreasonable, unjustly discriminatory or unduly preferential; and were effectively policed in order to guard against substitutions not authorized by the tariff. It, therefore, cancelled all Conference Rulings and Rule 76 of its tariff circular, dealing with milling in transit restrictions.

In abandoning the practice of expressing its views as to what would or would not constitute violations of the law in connection with transit privileges, the Commission placed upon the carrriers the duty of initiating their rates, regulations, and practices on transit by proper tariff provisions, and reminded them that in doing so they should proceed along the lines of a recognition of the responsibilities and liabilities imposed upon them by the Act, the Commission or the courts to take proper action in the event such rates, regulations, or practices were found to be in contravention of the statute, because of unjust discrimination, unreasonableness or undue preference, or where the matter of lawfulness of through routes and joint rates might be brought in issue through abuses of the substitution privilege.

Tariff Provisions to Prevent Abuse of Transit Privileges

Since the Commission rescinded its ruling in connection with transit and withdrew from making any further distinctions as to exactly how far the carriers could or could not go in granting and applying these privileges, the carriers in line with the Commission's expressions in this respect, formulated transit rules and published tariffs on various commodities, which in their opinions were in conformity with the law. In order to determine the feature of lawfulness definitely, every milling-in-transit arrangement must be carefully considered and disposed of on its own merits. In doing so, each commodity in itself, as well as every commercial and transportation feature which may affect the milling-in-transit arrangement, either in the lawfulness of its institution or in the legality of its application, must be carefully weighed and considered.

There are, however, certain general rules and practices which underlie the application of all milling-in-transit arrangements in order to guard against the possibility of misuses of these privileges by certain shippers and to insure compliance with those principles of the Act which govern the preservation of the through rate. The safeguards which have been placed around the transit tariffs of the carriers in order to prevent abuses of this nature are, briefly, as follows:

(1) **Persons Eligible to Receive Transit**—Only those persons or concerns located at the milling points who are definitely engaged in the trade as shippers, elevator operators, millers, or the like, are eligible to receive these privileges. Casual consignees or others not recognized as permanent receivers and shippers of grain may not be given the benefit of transit. This is in recognition of the instability of firms or individuals of this nature and the attendant fact that they could seldom, if ever, conform to the requirements as to the keeping of records, policing, etc.

(2) **Records**—The user of the transit privilege binds himself to keep accurate and complete records of all commodities and tonnage handled, and to submit to each inspection or policing by the carrier or its representative as may be deemed necessary in order to insure observance of the tariff. He must also, when requested, make statements, and, if required, affidavits as to the accuracy of such statements and records. The records required are detailed and comprehensive and must show:

(a) All grain and grain products that are handled.
(b) Point of shipment of each lot of grain.
(c) The destination of each consignment of grain products.
(d) The amount in pounds and kinds of grain received and product forwarded.
(e) When mixed shipments are forwarded, records must be kept so that they may be checked by the inspection bureau or other representative of the carrier granting the privilege.

In some cases separate records must be kept by the transit house indicating amount of grain:

(a) Received inbound via each mode of transport.
(b) Transferred from one transit house to another.
(c) Transit tonnage forwarded via each mode of transport.
(d) Disposed of locally.
(e) Local or non-transit tonnage forwarded.
(f) Changes in weight due to treatment or processing.
(g) On hand.

While these records are extensive and entail a great deal of additional clerical labor which the shipper is not ordinarily called upon to furnish in connection with shipments which are not stopped for transit, they are all necessary in order to enforce compliance with the tariffs. As transit is a privilege which is of distinct advantage to the shipper, he must take the bitter with the sweet and keep accurate and complete records if he expects to enjoy these privileges, even though they may cause him additional clerical labor and expense. The carriers inspection bureau will advise exactly what records and forms are necessary in any given instance.

(3) **Reports**—The user of the transit privilege must make periodical reports based on the records above referred to. These reports ordinarily call for statement of the tonnage received and forwarded each period, the

amount of shrinkage, and the amount of transit material on hand at the beginning and end of each period.

(4) **Invisible Loss in Weight**—Transit arrangements are based upon the theory that the incoming and outgoing transportation services, while distinct, are applied to a continuous shipment from point of origin to point of destination. It naturally follows then that if a car of wheat were to be shipped into a milling point and the identity of that particular car were to be preserved during the entire time it was passing through the process of being ground and loaded out again in the form of flour, there would be a natural loss or shrinkage in weight in the car of wheat which moved into the elevator as compared with the car of flour produced therefrom which moved out. If, therefore, the theory that the same car of grain moves out of the transit point as moved into it is to be observed, there must be an allowance made for this invisible loss in milling, otherwise local commodities to equal the amount of the shrinkage could be substituted and sent forward at a charge less than that to which they are entitled. For example, let it be assumed that a car of wheat weighing 60,000 pounds moves into the milling point and is malted. In this case the invisible loss or shrinkage on account of the malting process would be approximately 16% of the total weight of the whole grain, so that the outbound movement of the same commodity in its changed form would actually be but 50,400 pounds. If therefore, proper provision for the shrinkage were not made in the tariffs of the carriers and the outbound tonnage restricted accordingly, there would be nothing to prevent a shipper from forwarding 9,600 pounds additional of local malted grain and thus defeat the properly applicable tariff rate on such nontransit commodity. Tariffs provide for arbitrary percentage deductions to be made on the outbound movement according to the process through which the grain is passed. As an option to the arbitrary deductions provided for by the tariff, if shippers so elect they may determine and certify to the carrier, under oath if required, subject to verification by the inspection bureau having jurisdiction, the actual loss in weight due to the process of manufacture on their entire production, and deductions conforming to such certification will be applied in lieu of the specific arbitrary percentage ordinarily provided.

(5) **Recording and Cancellation of Billing**—On commodities on which transit is desired, the actual paid freight bills governing the inbound movements must be presented to the carriers for recording—usually within thirty days from date of issue.

The agent of the carrier then stamps or writes across the back thereof "Recorded for Transit," dates and signs the endorsement, and records the same. Then when shipments are forwarded from transit stations, unexpired inbound freight bills so recorded for transit, representative both in tonnage and kind of the commodity to be forwarded, must be surrendered to the carrier's agent and immediately cancelled. In this way a check is kept on the tonnage outbound and balanced against the tonnage actually received.

In the event a shipment should be made out of the transit point which would not equal the weight inbound (less proper deduction for invisible loss), a credit would be given the shipper by the carrier for the difference. Such excess weight or authorized credit would then be accepted on future shipments where the representative outbond tonnage, less deduction for loss, might be in excess of the inbound tonnage.

In addition, each day, representative freight bills covering inbound movements of grain or grain products disposed of locally or forwarded as non-transit tonnage must be surrendered to the carrier and cancelled, thus obviating the possibility of non-transit grain being later forwarded on this billing.

By a systematic surrender of freight bills in this manner a balance is maintained between the total uncancelled billing and transit credits and the total amounts of grain and grain products on hand at any one time.

(6) **Policing**—Because of the fact that the grain on which transit is allowed passes entirely out of the hands of the carriers, completely loses its individual identity by being mixed with other grains of the same kind and quality in the elevators and, in most instances, is converted into an entirely different commodity before it is again returned to the carrier, it is highly important that the records of the transit houses, as above referred to, be uniformly, accurately, and completely kept, and that proper cancellation of tonnage be made, or otherwise a myriad of opportunities to defeat the through rate will be presented. There could be a strong temptation on the part of some to take advantage of any opportunity of substitution which would permit them to obtain a lower rate than the tariff provides and thereby obtain an undue advantage over competitors bidding in the same market and observing the regular tariff rates and practices. In order to guard against the obtaining of such undue advantage in rates and the serious depletion of carrier's revenues by unauthorized substitutions of this nature, and to admit a thorough and impartial enjoyment of the privileges of transit, it is necessary for carriers to "police" the different arrangements, to see that they are applied strictly in accordance with the tariffs.

This is accomplished by means of bureaus designed for that purpose, who are experts in taking care of the highly technical matters which arise in connection with transit arrangements. They are the official representatives of the carriers and are maintained as a separate body by associations made up of the interested lines. They have the right, under the tariffs governing the transit privileges (which must be accepted by the users as condition precedent to receiving the benefit of transit), to enter the transit houses at all times and inspect the goods on hand, and to check the records of such concerns. They are also charged with the duty of cancelling freight bills covering tonnage in excess of transit grain and grain products in the transit houses, and of inspecting and supervising the records and billing in any other way that will conduce to the purpose for which they have been established.

While the duty to properly and effectively police transit privileges devolves primarily upon the carrier, and the interposition of no agency can relieve it of that responsibility, experience has demonstrated that because of negligence, inefficiency, pursuit of other duties, or the enormity of the task, every station agent cannot be relied upon adequately to assume the role of inspector, and that these bureaus can perform the policing far more effectively. At points where there is not sufficient business to justify the maintenance of an independent inspector, the agent of the carrier sometimes acts as agent for the inspection bureau, subject, however, to frequent check and constant supervision of the bureau.

The Commission has expressed itself as favoring these inspection bureaus because of their greater effectiveness in enforcing transit arrangements.

It has stated, however, that its approval of the inspection-bureau systems should not be understood as transferring the responsibility of proper policing from the carrier to the bureau, and that it recognizes the establishment of these bureaus as agents of the carriers in the discharge of a duty imposed upon them by law.

Resume—To sum up, the primary safeguards which have been placed around the proper enforcement of milling in transit arrangements by the various tariffs of the carriers, in order to insure compliance with the laws, are as follows:

(1) Restriction of persons eligible to receive transit.
(2) Keeping of proper records.
(3) Rendering of periodical reports.
(4) Making of allowance for invisible loss in weight.
(5) Recording and cancellation of billing.
(6) Policing.

These restrictions are as necessary to the safe and orderly enforcement of the transit arrangement as is the need for the privilege in connection with a given commodity in the first instance. They are so important in fact, that a transit tariff cannot be justified, or the privilege granted thereunder rendered legally operative, unless the nature of the commodity and of the transit arrangement contemplated is such as to admit a thorough application of these restrictions, to the end that the very purpose for which the privilege is established may not be defeated and thereby, under the guise of official tariff sanction, bring about a violation of the law.

Time Limits

Carrier transit tariffs ordinarily limit the time which a shipment may be held at the transit point for the purpose for which it is stopped. This time limit is usually twelve months. The length of time the transit tonnage may stay at the transit point is a matter of judgment with the carrier granting the privilege. It may be as short a period as three months or as long as six years. In *Alcoholic Liquors, 6 Year Stop in Transit for Aging,* (305 I.C.C. 798), the Commission said:

> "Proposed period of six years for stopping in transit for aging of alcoholic liquors found reasonable. While transit periods extending beyond one year have long been regarded by the Commission as prima facie unreasonable, longer periods have been approved where there was a general demand therefor. Time limitations may vary with the commodity, the locality, manner of conducting business, and changing competitive relations among carriers as well as shippers; transit provisions which take into account fact that alcoholic liquors are usually aged from four to six years rest on a sound basis; moreover proposal would enable respondents to meet private truck competition."

Transit time limitations are usually measured from the date of the freight bill covering the inbound shipment to the transit point. When the time limit expires the transit privilege ceases and all shipments remaining in the transit account beyond the time limit are considered separate and distinct shipments to and from the transit point.

To assist shippers in situations where something unforeseen develops making it impossible for the shipper to reship from the transit point prior to the expiration of the time limit the carriers provide in their transit tariffs a rule under which the shipper may be granted an extension of time. Requests for an extension of time must be made prior to the expiration of the original time period. Tariffs cannot be given a retroactive effect and once the time limit in connection with a freight bill has actually expired a time limit extension may not make the freight bill good again for transit. Carriers may make a charge for the time limit extension which charge will be in addition to any transit charge that may be applicable.

Application of Rates

The application of rates on shipments accorded a transit privilege will be provided in the tariff naming the privilege. Generally, carrier tariffs provide for the application of the through rate from point of origin to ultimate destination on either the inbound or the outbound commodity whichever is higher. The term through rate may mean either a one figure joint through rate or a combination of rates. Very frequently carriers establish special reshipping or proportional rates from the transit point which added to the local or in some cases proportional rate paid into the transit point produces a total charge equal to the through rate or where no through rate is provided somewhat lower than the combination of local rates over the transit point. In all cases the transit tariff should be checked very carefully to determine the exact application of rates. The one factor that should always be remembered in connection with the rate to apply is that since transit is based upon the fiction of a through continuous movement from origin to destination, the rate in effect at the time for shipment begins is the legal rate to apply.

Where the transit tariff provides for the application of the through rate from point of origin to ultimate destination and the only available rate is a combination of local, joint or proportional rates there is a tendency to consider only that combination made over the transit point. In other words, a combination of the rate to the transit point and the rate from the transit point to the destination. Actually the rate to apply is the lowest combination of rates that can be computed via the route of movement. The point over which the combination is made does not necessarily have to be the transit point.

In applying rates on transit shipments little difficulty is encountered in connection with the rate to the transit point, it is the rate or rates to be applied from the transit station to the transit destination that cause the trouble. This is particularly true in cases of "split billing" under "balance out" rates. Under the "balance out" arrangement the shipment from the transit station is billed at the balance of the through rate. This balance is the difference between the through rate from point of origin to ultimate destination and the rate from origin to the transit station. Where only one origin is involved the outbound billing is comparatively simple. Under the split billing arrangement, carriers permit the use of tonnage and the surrender of billing from several origins. In such cases it is necessary to break down the total weight on the basis of the actual weights from each of the respective points of origin. This requires the checking of the through rate from each of the origin points to the ultimate destination, and the applying on each of the respective weights the difference between the amount paid in on that portion and the through rate from that particular point of origin. In some cases transit tariffs limit the split billing of an outbound car to a certain number of origins such as three or five. However, where no limitation is placed upon the split billing an outbound car may consist of tonnage from ten or more origins.

The total tonnage that may be shipped from the transit station is based upon the actual weight of the inbound shipment and not upon the minimum weight attached to the inbound rate. Quite frequently the weight of the shipment from the transit station to the transit destination is less than the required minimum weight. In cases such as this it is extremely important to determine what provision is made in the tariff for the handling of this deficit in weight. In the majority of cases the deficit in weight is chargeable at the local rate from the transit point on the outbound product. There are some cases however where tariffs provide for the surrender of transit billing to cover the deficit in weight. There are many transit arrangements where weight is added at the transit point as in the case of fabrication of iron and steel resulting in a greater weight outbound from the transit station than arrived inbound. In such cases the maximum which may be forwarded under the transit arrangement is the total of the inbound tonnage and any additional weight is chargeable at the local rate from the transit point to the destination.

The same situation is encountered with respect to non-transit tonnage which is mixed with transit tonnage at time of forwarding from the transit point. Tariffs generally provide that non-transit tonnage consisting of tonnage purchased locally, or tonnage on which the time limit has expired, or tonnage which moved into the transit point via some other medium of transportation, may be shipped in mixed carloads with transit tonnage. The rate to apply on the non-transit tonnage is the local rate from the transit point to destination. In such cases it is necessary to determine whether or not the non-transit tonnage may be used to make up the required minimum weight.

As a general rule transit tariffs provide that the transit station must be directly intermediate between point of origin and ultimate destination. Quite frequently, however, transit arrangements are established at points that are not directly intermediate and which involve an out of route movement or a back haul. In some cases such arrangements are permitted without additional charge and in others a charge is made, usually on a mileage basis.

Territorial limitations specified in a transit tariff must be strictly observed. If a transit tariff states that the privilege is applicable only on traffic originating in a certain specified territory then such privilege cannot be applied on traffic originating at some point outside of the specified area. Furthermore, it is not permissible to "tack" to a territorial limitation.

Many transit tariffs provide, under the application of rates, that the minimum through rate to be applied shall be:

1. The flat rate from origin to destination, or
2. The flat rate from origin to transit point, or
3. The flat rate from transit point to destination, whichever is

highest, on the commodity received at the transit point or the commodity forwarded from the transit point, whichever is higher, in effect on date of shipment from point of origin.

The Interstate Commerce Commerce Commission has upheld the publication of the three-way rule in numerous cases. In *Grain and Grain Products,* (205 I.C.C. 301), the Commission said:

> "Whether the three-way rule was originally established as a penalty for out of line service or to prevent breaking down the rate structure is not clear. It serves both purposes in that it usually is encountered on routes over which fourth-section relief has been obtained. If, for example, a shipment from a more distant and lower rated point is stopped in transit at an intermediate but higher-rated point on a circuitous route, the transit operator at the higher-rated point could, by switching billing, always move his local shipments from the higher-rated point at the balance of the through rate from the more distant point unless restricted by the three-way rule. The application of the higher rate from the intermediate point therefore not only protects the rate structure but also compensates the carrier for the extra service performed. It also guarantees the carrier against refunding any of its revenue to the transit point when the through rate is less than the rate to the transit point."

In *Coarse Grains for Feeding in WTL Territory,* (266 I.C.C. 773), the Commission said:

> "The Commission has recognized the inherent justice of the three-way rule and found it to be a necessary and reasonable transit requirement."

Also, in *South Texas Cotton Oil Co., v. A. & S. R.R.,* (297 I.C.C. 767), the Commission ruled:

> "It has long been established that the term 'through rate' as used in the 3-way rule in transit tariffs may be either a single factor rate or a combination rate."

Motor Carrier Transit Privileges

While motor carriers have established some transit privileges they are relatively few in comparison with those of the railroads. It would seem, however, that with the establishment of more and more through routes and joint through rates more transit privileges will be requested by shippers and receivers and established by the carriers. Furthermore the principles that have been established in connection with rail privileges will undoubtedly hold true as far as motor carrier privileges are concerned.

In *Stoppage in Transit, Central Territory,* (51 M.C.C. 25), the Commission had the following to say regarding motor carrier transit:

> "Transit service is of a special nature and, except where it is universally available and necessary in respect of a particular commodity, should not be required by the Commission except upon a clear showing that failure to provide it results in exaction of unreasonable rates or in undue prejudice."

Also on the subject of transit in connection with motor carrier shipments they said, in *Transit Privileges at Ransom, W. V.*, (63 M.C.C. 660):

> "A transit privilege rests upon a legal fiction that the transportation services to and beyond the transit point constitute a continuous shipment of the identical article from origin to destination. In motor-carrier operation, the privilege upon which the fiction is predicated must be in consonance with the carrier's operating authority. The legal fiction only forms a basis for assessment of more favorable rates than would obtain if the inbound and outbound movements to transit points were treated as separate shipments."

Elevation as a Transportation Necessity

Elevation is more than a privilege of value to the shipper; it is also a transportation necessity which the law says the carrier must furnish as part of their business.

Public Law 95-473 makes it the duty of every common carrier subject to the Act to furnish transportation upon reasonable request therefore. In defining the term transportation, the law provides that it shall include "services relating to that movement including receipt, delivery, elevation, transfer in transit ****" Elevation has, therefore, been deemed an inherent part of transportation and the carriers are required by law to furnish such service as a part of the transportation which they hold themselves out to perform.

Dual Nature of Elevation

In speaking of elevation as a transit privilege, it was developed from statements of the Commission that it is not incumbent upon a carrier to furnish such services, but that they may be granted or withheld so long as no discrimination results thereby. It would, therefore, seem as though there exists a conflict between the actual text of the law and the Commission's findings in respect to the features of elevation. When it is understood, however, that from a traffic standpoint there are two separate and distinct kinds of elevation, and that the requirements of the law apply to one and the granting of special privileges to the other, we arrive at a better appreciation of the confusion and misapprehensions that may be brought about through the lack of a proper understanding of the dual nature of this term.

In speaking of this difference, the Commission stated, in the *Matter of Elevation Allowances at Points Located upon the Missouri, Mississippi and Ohio Rivers and on the Great Lakes,* (24 I.C.C. 197, 199), that:

> "There are two kinds of elevation, one of which may be termed *transportation elevation,* consisting of the passing of the grain through an elevator for the purpose of transferring it from car to car and obtaining its weight, and *commercial elevation,* which involves various processes in the treatment of the grain itself, like cleaning, mixing, clipping, drying, etc. The first sort of elevation is an incident to the transportation of the grain, the second to the merchandising of the grain."

Also *In the Matter of Allowances to Elevators by the Union Pacific Railroad Company,* (12 I.C.C. 85, 87), the Commission made the following differentiation in the services:

> "Elevation, as commonly understood among elevator men and among buyers and sllers of grain, signifies the unloading of grain from cars, or from grain-carrying vessels, into a grain elevator and loading it out again after storage for a period of not to exceed ten days. The 'treatment,' or grading, cleaning, and clipping of grain is not properly a part of 'elevation' as the word is strictly used, and the retention of grain in an elevator beyond the period of ten days becomes storage and is not elevation. It is in this sense that the word is used in the amended statute."

In the absence of any definite rules in this respect, the circumstances surrounding the case at hand must be taken into consideration. If, therefore, a given situation fairly and equitably meets in an affirmative manner the question: "Is this elevation incident and absolutely necessary to the transportation which the carrier is required by law to furnish," it may be rightfully considered a "transportation elevation." If it does not fairly answer this question, then it should be considered as a "commercial elevation" of particular value to the shipper.

Carriers May Furnish Commercial Elevation

At the time the grain is passing through the required "transportation elevation" at a carrier's elevator, it may also be put through other processes of a "commercial" nature which are entirely apart from the service of carriage, such as, cleaning, milling, grading, etc. While these additional privileges of commercial elevation may be, and quite frequently are, furnished by carriers, they are not required by law to do so.

Elevation of this nature may be furnished by a carrier only when specifically provided for in duly published tariffs, open to all shippers alike, and the charges therefore, in line with the underlying principles of the Interstate Commerce Act, must be just and reasonable. In this connection, the Commission, in the case of *Traffic Bureau, Merchants' Exchange of St. Louis v. C.B. & Q.R. Co.*, (14 I.C.C. 317, 331), stated:

> "We do not now hold that commercial elevation may not properly be furnished by a railroad, but we do hold that such elevation must be charged for at what it is reasonably worth."

In *Re Elevation Allowances*, (24 I.C.C 197, 203), the following appears:

> "*** We must not only prohibit the railroad from rendering for the shipper at its own elevator, free, any service beyond transportation elevation proper. We must go further. We must determine what is a just charge for these commercial operations and insist that the railroad elevator, if it performs the operations, shall charge not less than the sums found reasonable."

It should be understood, however, that while a carrier *may furnish* such commercial elevation, it is under no legal obligation to do so. The following expression of the Commission, as contained in the case of *Grain Elevation Allowances at Kansas City, Mo., and Other Points*, (34 I.C.C. 142, 447), brings this out very emphatically:

> "It is clear that the elevation *required* of carriers by Section 1, and for which, if rendered by an elevator operator, they may make an allowance under Section 15, is such elevation as is reasonably necessary to the transportation involved; and that carriers are *not* required by the Act to furnish elevation desired by the shipper for a commercial reason, but may permit stops in transit for such elevation, a transit service which, like all others, is subject to regulation by the Commission."

CHAPTER 10

RELEASED AND ACTUAL VALUE RATES

While it is not our purpose here to explore in depth the subject of carrier liability it is necessary, in order to adequately describe released and actual value rates, to consider the liability of carriers thereunder. On page 159 we have reproduced Items 88140 and 88150 of the National Motor Freight Classification. These items provide ratings on Glassware, NOI. Item 88140 names released ratings on Glassware having a released value not exceeding 35 cents per pound to $7.50 per pound and Item 88150 names actual value ratings on Glassware valued at from under 35 cents per pound to $5.00 per pound. In each case the shipper must place a declaration on the bill of lading specifying the released or actual value of the Glassware being tendered for shipment. The items are quite similar and the declarations of value are rather similar but, actually, there is a great difference between the two items.

In order to properly understand the application of rates and ratings predicated upon a declaration of value by the shipper as well as the significance of such declarations in the event the shipment is lost or damaged it is necessary to look first to the regulatory law. Section 11707(a) of P.L. 95-473 makes common carriers (except water carriers) liable for the actual loss or injury to the property caused by the carrier. Further, it provides in subsection (c) that:

> "(1) A common carrier may not limit or be exempt from liability imposed under subsection (a) of this section except as provided in this subsection. A limitation of liability or of the amount of recovery or representation or agreement in a receipt, bill of lading, contract, rule, or tariff filed with the Commission in violation of this section is void.
>
> "(2) If loss or injury to property occurs while it is in the custody of a water carrier, the liability of that carrier is determined by its bill of lading and the law applicable to water transportation. The liability of the initial or delivering carrier is the same as the liability of the water carrier.
>
> "(3) A common carrier of passengers may limit its liability under its passenger rate for loss or injury of baggage carried on passenger trains, boats, or motor vehicles, or on trains, or boats, or motor vehicles carrying passengers.

"(4) A common carrier may limits its liability for loss or injury of property transported under section 10730 of this title.

"(d) A civil action under this section may be brought against a delivering carrier in a district court of the United States or in a state court. Trial, if the action is brought in a district court of the United States is in a judicial district, and if in a state court, is in a state, through which the defendant carrier operates a railroad or route.

"(e) A carrier may not provide by rule, contract, or otherwise, a period of less than 9 months for filing a claim against it under this section and a period of less than 2 years for bringing a civil action against it under this section. The period for bringing a civil action is computed from the date that person receives written notice from the carrier that it has disallowed any part of the claim specified in the notice."

From the above it will be noted that a carrier may limit its liability under section 10730 which reads as follows:

The Interstate Commerce Commission may require or authorize a carrier providing transportation or service subject to its jurisdiction under subchapter I, II, or IV of chapter 105 of this title, to establish rates for transportation of property under which the liability of the carrier for that property is limited to a value established by written declaration of the shipper, or by a written agreement, when that value would be reasonable under the circumstances surrounding the transportation. A rate may be made applicable under this section to livestock only if the livestock is valuable chiefly for breeding, racing, show purposes, or other special uses. A tariff filed with the Commission under subchapter IV of this chapter shall refer specifically to the action of the Commission under this section."

It will be noted from the above that the Commission is empowered to authorize the publication of rates dependent upon a value declared in writing by the shipper or agreed upon in writing as the released value of the property and that when such rates are published, specific reference to the order of the Commission authorizing such rates shall be shown. Examining Item 88140 of the NMFC it will be noted that it states "classes herein based on released value have been authorized by the Interstate Commerce Commission in *Released Rates Order No. MC-607 of April 15, 1965.* ***" Looking next at Item 88150 it will be seem that there is no reference to an order of the Commission; furthermore, Item 88150 provides a scale of "actual" values whereas Item 88140 lists a scale of "released" values.

It is apparent therefore that while released rates must be specifically authorized by the I.C.C. there is no such requirement as far as actual value rates are concerned. Furthermore, a careful reading of the declaration which must be placed on the bill of lading in each instance will

NATIONAL MOTOR FREIGHT CLASSIFICATION 100-E

Item	ARTICLES	CLASSES LTL	CLASSES TL	MW
	GLASSWARE GROUP: subject to item 87500			
88070	**Glass-ceramic Ware,** see Note, item 88074, **or Laminated Glassware,** cooking or serving; see Notes, item 88072 and 88076, in boxes:			
Sub 1	Released value not exceeding 35 cents per pound	70	37½	28.2
Sub 2	Released value exceeding 35 cents but not exceeding $1.50 per pound	85	55	24.2
Sub 3	Released value exceeding $1.50 but not exceeding $2.00 per pound	100	70	20.2
Sub 4	Released value exceeding $2.00 but not exceeding $3.75 per pound	150	85	20.2
Sub 5	Released value exceeding $3.75 but not exceeding $5.00 per pound	200	100	20.2
88072	NOTE—The released value must be entered on the shipping order and bill of lading in the following form: *"The agreed or declared value of the property is hereby specifically stated by the shipper to be not exceeding—— per pound."* (Classes herein based on released value have been authorized by the Interstate Commerce Commission in Released Rates Order No. MC-545 of August 29, 1963 as amended May 15, 1964 and February 13, 1973, subject to complaint or suspension.) (See page 3 for state authorities.)			
88074	NOTE—Applies on articles of cooking or table serving utensils (except straight shipments of such articles equipped with electrical heating units) or dishes made of a ceramic that consists of glass that has been converted into crystalline ceramics by the use of nucleating agents, heat treatment and kiln ceraming. The resulting material is a glass derivative opaque ceramic characterized by greater strength and hardness than the parent glass and has greater abrasion resistance.			
88076	NOTE—Released classes will also apply on glass-ceramic ware or laminated glassware equipped with candles, covers, handles, metal or plastic fixtures, serving baskets or table type holders or stands not in excess of number required to equip each retail unit of sale. Glass-ceramic ware or laminated glassware equipped with electric heating units may be included in the shipment at the classes applicable to glass-ceramic ware or laminated glassware but not in excess of 25 percent of the total weight of shipment on which charges are assessed			
88080	**Glass,** single layer, cut to shape for goggles, sunglasses, helmets or body protective shields, other than optically ground for vision corrective purposes, in boxes	85	55	30.2
88100	**Glasses, gauge, solid glass, not bent nor curved, in boxes**	100	70	24.2
88120	**Glassware,** laboratory, **including Beakers, Bulbs, Evaporators, Pipettes** NOI, **Test Tubes or Worms,** in barrels or boxes or in Package 1344	100	70	18.2
88140	**Glassware,** NOI, see Notes, items 88142, 88143, 88154, 88156, 88166 and 88172 in barrels, boxes or in Packages 183, 196, 563, 1174, 1346, 1395, 1422, 2089, 2126 and 2127:			
Sub 1	Released to a value not exceeding 35 cents per pound	70	37½	28.2
Sub 2	Released to a value exceedng 35 cents per pound but not exceeding $1.50 per pound	85	55	24.2
Sub 3	Released to a value exceeding $1.50 per pound but not exceeding $2.00 per pound	100	70	20.2
Sub 4	Released to a value exceeding $2.00 per pound but not exceeding $3.75 per pound	150	85	20.2
Sub 5	Released to a value exceeding $3.75 per pound but not exceeding $5.00 per pound	200	100	20.2
Sub 6	Released to a value exceeding $5.00 per pound but not exceeding $7.50 per pound	250	125	20.2
88142	NOTE—The released value must be entered on the shipping order and bill of lading in the following form. *"The agreed or declared value of the property is hereby specifically stated by the shipper to be not exceeding—— per pound."* (Classes herein based on released value have been authorized by the Interstate Commerce Commission in Released Rates Order No. MC-607 of April 15, 1965, as amended March 5, 1974, subject to complaint or suspension.) (See page 3 of classification for state authorities.)			
88143	NOTE—If the shipper declines or fails to declare the value or agree to a released value in writing the shipment will not be accepted, but if the shipment is inadvertently accepted, charges will initially be assessed on basis of the class for the highest valuation provided. Upon proof of the lower actual value, freight charges will be adjusted to the basis of the class or rate applicable in connection with such actual value. In no instance will carriers' liability exceed the highest value for which classes are provided.			
88150	**Glassware,** NOI, in barrels or boxes, see Notes, item 88152, 88154, 88156 and 88166, or in Packages 183, 196, 563, 1174, 1346, 1395, 1422, 2089, 2126 or 2127:			
Sub 1	Actual value not exceeding 35 cents per pound	70	37½	28.2
Sub 2	Actual value exceeding 35 cents per pound, but not exceeding $1.50 per pound	85	55	24.2
Sub 3	Actual value exceeding $1.50 per pound, but not exceeding $2.00 per pound	100	70	20.2
Sub 4	Actual value exceeding $2.00 per pound, but not exceeding $3.75 per pound	150	85	20.2
Sub 5	Actual value exceeding $3.75 per pound, but not exceeding $5.00 per pound	200	100	20.2
Sub 6	Except as provided in items 88140 and 88172, if actual value exceeds $5.00 per pound, or if shipper declines to declare actual value	Not	Taken	

For explanation of abbreviations and reference marks, see last page of this tariff.

disclose that in one case the shipper must specify the "agreed or declared" value of the property but in the other they must declare the "actual value" in accordance with the scale of values shown in the item.

In the case of released rates there must be a choice. The shipper must be given the opportunity to choose either a higher rate with little or no limitation of liability or a lower rate in return for which he agrees to limit the carrier's liability to a specified amount. The courts have consistently held that a limitation of liability can be sustained only where a choice of rates has been given and a limitation of liability made the basis of the reduced rate.

For the purposes of illustration let us assume that a manufacturer of glassware had a 100 pound shipment of glass goblets valued at $7.00 per pound. In order to obtain the lower rating he indicated on the bill of lading that "the agreed or declared value *** to be not exceeding $5.00 per pound." If the shipment was lost or completely destroyed in transit the shipper could recover a total of only $500.00 (100 pounds x $5.00 per pound)despite the fact that the actual value of the merchandise was $700.00. Furthermore, there was no misbilling or misrepresentation by virtue of the declaration of a value lower than the actual value.

If, however, the shipment of 100 pounds had an actual value of $4.00 per pound and the shipping clerk preparing the bill of lading inadvertently declared the actual value to be "in excess of 35 cents per pound but not exceeding $1.50 per pound" and the shipment was lost or destroyed the result would be different.

Under the law (Section 11707 of P.L. 95-473) the shipper is entitled to recover his "full actual loss" which in this case would be $400.00. It would, of course, be evident to the carrier, upon settlement of the claim, that charges were assessed on the wrong basis and a balance due bill would be issued correcting the charges to the rating based on the higher actual value.

If it could be shown that the shipper had knowingly misstated the value of the merchandise so as to obtain the lower rating they could be prosecuted for violation of the Elkins Act. Furthermore, there have been cases in which the courts have ruled that in such situations the shipper, by virtue of such misrepresentation, has forfeited his right to recover for the loss. In *Hart v. Pennsylvania R.R.,* (112 U.S. 331), the court ruled:

The following citations will serve to indicate the thinking of the Commission with regard to released value rates:

"If the shipper is guilty of fraud or imposition, by mis-representing the nature or value of the articles, he destroys his claim to indemnity, because he has attempted to deprive the carrier of the right to be compensated in proportion to the value of the articles and the consequent risk assumed, and what he has done has tended to lessen the vigilance the carrier would otherwise have bestowed. *** The compensation for carriage is based on that value. The shipper is estopped from saying that the value is greater. The articles have no greater value, for the purpose of the contract of transportation, between the parties to that contract."

When susceptibility to loss or damage is comparatively high and the wide range in value of the commodity makes the amount of any claim that may arise difficult to estimate, the carrier is at a disadvantage unless it is permitted to base its liability and its charges on a declaration of value obtained in advance from the shipper. *Released Ratings on Engines,* (287 I.C.C. 419).

Although from a commercial standpoint there was no difference in medicines shipped whether their value was released or not, there was a distinct difference from a transportation standpoint by reason of shipper's declaration of value. Following *American Home Foods Case,* (303 I.C.C. 655), released and unreleased rates on same commodities in uniform classification should be considered as separate and distinct items, especially where the released rate item was covered by Commission's order in effect during period of movement. Released class rate claimed was not displaced by higher commodity rate not contingent upon declared or released value and was therefore applicable. *Upjohn Co. v. Pennsylvania R.R.,* (306 I.C.C. 325).

From a commercial standpoint there was no difference in the commodity shipped whether its value was released or not released, but there was a distinct difference from a transportation standpoint by reason of shippers delcaration as to value. An exceptions rating subject to released value would not have removed unreleased rating from the classification, and conversely the unreleased exceptions rating did not remove released ratings therefrom. Released and unreleased ratings are separate and distinct items from a transportation standpoint. *Dow Chemical Co. v. C. & O. Ry.,* (306 I.C.C. 403).

The Commission's discretionary power under Section 20(11) (now Section 10730) to authorize released rates is exercised sparingly, since it involves an exception to the general policy against any limitation of liability by a common carrier; and general principles governing establishment of such rates as enunciated in 93 I.C.C. 90, require a showing, before partial exemption from full liability may be approved, that susceptibility of the traffic to loss and damage is comparatively high and that the wide range in value of the commodity makes the amount of any claim that may arise difficult to estimate. *People's Express Co.*, (311 I.C.C. 515).

CHAPTER 11

WAREHOUSING AND DISTRIBUTION

General

We now come to a study of an important commercial service which is inseparably linked with commerce, and is known as warehousing. While "warehousing" implies storage, "warehousing" is more than storing. It is storage which is not contemplated by any condition incorporated in the uniform bill of lading. Such services are performed throughout the country by commercial warehousemen under and pursuant to their private contracts, arrangements and dealings with patrons of warehouses. Many manufacturers establish and operate their own private warehouses. Our discussion here, however, will be confined primarily to the commercial or public warehouse. Such services usually result from solicitation by the party who desires to perform the services and are voluntarily performed by him for hire or reward.

Storage by carriers, of freight which they transport for hire, is a service incidental to transportation. Such storage is defined in the uniform bill of lading and results when the consignee does not accept delivery of his shipments promptly. The carrier is then compelled to store the goods at rates sufficiently high to encourage the shipper to remove his goods as quickly as possible in order to release the space or the car. In other words, one is a voluntary solicited service of a commercial warehouse concern engaged in trade activities, referred to as commercial warehousing; the other is the voluntary incidental service of the common carrier.

Commercial Warehouses

Warehousing has a written history extending back into ancient Biblical times. The commercial warehouse industry is an important part of the transportation and distribution system. It has always been necessary for a business, dealing in large quantities of merchandise, either to provide and operate its own private warehouse facilities or to use those provided by public warehousemen for the storage and distribution of its goods. While warehousing is an independent industry, it has a close relationship with transportation, banking and insurance.

Businesses engaging in commerce on a large scale frequently depend upon banks to help them finance their operations. One method of obtaining the necessary financing is to have the warehouse submit the receipt for goods stored to the bank. The warehouse receipt serves as collateral for a loan to the depositor of the goods. The actual service performed by the warehouse is the same and the only difference is that title to the goods rests with the bank until such time as the loan is repaid.

The warehousing industry today is divided and subdivided within itself into warehouses specializing in storing and handling a single commodity or groups of commodities. For example: there is the warehousing of grain in elevators; the warehousing of perishable foods in cold storage warehouses; the warehouses catering solely to household goods and furniture; the warehouses handling exclusively wool, cotton, tobacco and numerous other specific commodities; the warehouses storing goods in custody of the United States Customs and Internal Revenue Departments, known as Bonded Warehouses; and finally the merchandise warehouses, handling and distributing manufactured and semi-manufactured merchandise of every description. Each of the foregoing kinds of warehouse is a business in itself. The merchandising warehouse is by far the largest part of the warehousing industry as a whole, so our present study of "Warehousing and Distribution" will deal primarily with that type of warehouse.

Functions of the Merchandise Warehouse

The functions of the merchandise warehouse are numerous. In addition to the primary function of storing there are attendant functions such as: liability for the goods, inspection and grading, sorting, labeling and packing, conditioning of goods, re-shipment and transfer, financing of depositors, exhibiting and selling, distribution, pool cars, and trucking. Warehouse services are ordinarily used for:

A—**The storage of seasonal commodities or manufactured articles because of over-production.** Certain commodities are not grown or produced at the same time they are consumed or in demand. Storage makes these commodities procurable regardless of the season thereby preventing much waste. Many manufactured articles produced in large quantities during certain seasons because of climatic conditions, or because manufacturing processes are possible or are more economical under certain conditions or because of lack of space at point of manufacture, are also stored and shipped out according to current demand. These commodities are usually shipped in and out of the warehouse in carload quantities and are known as wholesale storage.

B—**Commodities stored for distribution**—Under present merchandise marketing much general merchandise is sold in small lots. Consumers, retailers, wholesalers and jobbers have forced this method of marketing by eliminating the old practice of "stocking-up" several times a year and adopting the practice of reducing stocks to a minimum so that they may have a quick turnover which promotes greater profits. This method of marketing has resulted in buyers purchasing smaller quantities of merchandise and placing orders more frequently. The manufacturers and sellers of merchandise, in order to meet competition and make prompt delivery of goods, warehouse their merchandise in many cities and towns throughout the country. The manufacturer or seller ships his merchandise in volume quantities to the warehouse, located at strategic points, at intervals depending on the market demand. This merchandise is then stored and the warehouse makes distribution according to orders received. Local shipments are called for by customers or delivered to them by the warehouse trucks or local cartage companies. Shipments to nearby towns are shipped by the warehouse in less-than-carload or less-than-truckload quantities. The rapid growth of this warehouse function has proven the value of the warehouseman as an important factor in the economical and efficient distribution of merchandise.

C—**Distribution of Pooled Shipments**—This function of the warehouseman permits tradesmen to take advantage of lower rates in carload and truckload lots. Volume lots of one kind or of assorted kinds of merchandise, purchased in small lots by various customers, are shipped to a centrally located warehouse for distribution ot the various small-lot purchasers of the merchandise, the purchasers thereby obtaining the benefit of lower rates. This function is similar to the function described in "B" above except that the merchandise generally does not go into storage, each small-lot being distributed immediately upon arrival of the car at the warehouse. This pooling operation is extensively used for distributing manufactured goods of all sorts.

Warehouse Receipt

In addition to the principal functions outlined above a warehouseman performs numerous incidental services. One of these is the issuance of warehouse receipts of which there are two kinds, negotiable and non-negotiable. The receipts most widely used throughout the United States are those adopted by the American Warehousemen's Association in 1968, copies of which are reproduced on pages 000 and 000.

The following commentary on the Negotiable Warehouse Receipt was prepared by the American Warehousemen's Association.

"The Negotiable Warehouse Receipt form, size 8½" x 11", conforms in every respect with the requirements of the Uniform Commercial Code and the Uniform Warehouse Receipt Act (Louisiana), wherein are listed the data that must be embodied within a warehouse receipt. A Negotiable Warehouse Receipt is negotiable, as distinguished from a non-negotiable document, by virtue of its wording as to whom delivery will be made. The AWA Standard Negotiable Warehouse Receipt is printed on green safety paper and is, purposely, entirely different in appearance from the Non-Negotiable Warehouse Receipt as an added means of distinguishing it from the latter document. A warehouseman issuing a Negotiable Warehouse Receipt must, of course, be familiar with the Uniform Commercial Code or the Uniform Warehouse Receipts Act (Louisiana), as they apply to the additional obligations he incurs in connection with his issuance of such a receipt.

"The face of the Negotiable Warehouse Receipt also has space for entering partial deliveries and space for endorsement. As required by law, Negotiable Warehouse Receipts must be surrendered for cancellation or notation of partial deliveries, and the warehouseman must cancel the document or conspicuously note the partial delivery thereon, or be liable to any party to whom the document is duly negotiated. Under the Uniform Warehouse Receipts Act, delivery of goods by any officer, agent or servant of a warehouseman without obtaining the negotiable receipt is a criminal offense. While the Uniform Commercial Code does not contain criminal provisions, the criminal codes of most states provide for fine or imprisonment for delivery of goods without surrender of the negotiable receipt.

"On the reverse side of the AWA Negotiable Warehouse Receipt form is the full text of the Standard Contract Terms and Conditions for Merchandise Warehousemen as approved and promulgated by the American Warehousemen's Association, October, 1968.

"In the left center of the face of the Receipt there is a statement reading: 'The property covered by this receipt has NOT been insured by this company for the benefit of the depositor against fire or any other casualty.' This statement appears also on the Non-Negotiable Warehouse Receipt. In some few states, warehousemen may be required by law to insure certain agricultural products. Warehousemen operating in those states will, of course, wish to check their position, legally, before having the 'not insured' statement printed on their receipts.

"The second copy of the Negotiable Warehouse Receipt is, primarily, a blank piece of yellow paper containing only the warehouse company's name, the consecutive number, and the heading 'Office Copy.' A carbon copy of the original Negotiable Receipt will show all the information typed on the original but none of the printed text of the original. The Office Copy is intentionally designed in this manner so that no duplicate containing the full wording of the original would be available to anyone who might be inclined to represent such a duplicate as a negotiable document."

The complete text of the contract terms and conditions, which are printed on the back of both the negotiable and the non-negotiable receipts, has been reproduced and will be found at the end of this chapter. In addition, the pertinent provisions of the Uniform Commercial Code which is applicable in all states with the exception of Louisiana, have also been reproduced at the end of this chapter.

The form of warehouse receipt may differ in actual use, principally in the manner in which title to the goods may pass from party to party, and the manner in which delivery of the goods may be effected.

The negotiable receipt passes from person to person by endorsement. The person holding the negotiable receipt has title to the merchandise and delivery of the merchandise cannot be made without surrender of the receipt properly endorsed.

The non-negotiable warehouse receipt is merely evidence that the merchandise has been placed in the hands of the warehouseman and surrender of the non-negotiable receipt is not necessary in order to obtain delivery of the merchandise.

Warehouse Services

Inspection—Before issuing his warehouse receipt the warehouseman makes a careful examination of the merchandise upon arrival to verify the quantity—weight, volume or gallonage, count of units, and the condition of the containers (if in packages), or of the warehoused goods, marking, tagging and labeling of the merchandise is classified according to: type of containers; difficulty of stowing, or tiering and piling; rigidity; fragility; excess value; hazard of fire; liability to cause damage; excess care required; liability to freeze; and many other factors. This inspection is not only a verification of quantity and condition of the merchandise received but it is also an appraisal of value and a certification of grade and quality of merchandise.

Sorting—The service of sorting is performed by the warehouseman in the case of carload or truckload quantities of merchandise consisting of goods of assorted brands, articles of different sizes or models, agricultural staples received in bulk, shipments containing the merchandise of several owners, etc. Only by sorting merchandise received can articles be segregated for proper storing and only by this service can they be recorded so as to be readily available for delivery.

Labeling—To maintain the identity and to meet certain legal requirements of warehoused goods, marking, tagging, and labeling of the

AWA Seal of Security

AMERICAN WAREHOUSE COMPANY
A PUBLIC WAREHOUSE
2121 AMERICAN AVENUE • AMERICA

Date of Issue__________ Consecutive No.__________

THIS IS TO CERTIFY that we have received in Storage Warehouse__________

situated at__________

for the account of__________

in apparent good order, except as noted hereon (**contents, condition and quality unknown**) the following described property, subject to all the terms and conditions contained herein and on the reverse hereof, such property to be delivered to (His) (Their) (Its) order, upon payment of all storage, handling and other charges and the surrender of this Warehouse Receipt properly endorsed,

LOT NO.	QUANTITY	SAID TO BE OR CONTAIN	STORAGE PER MONTH		HANDLING IN AND OUT	
			RATE	PER	RATE	PER
		NEGOTIABLE Quantities subject to deliveries noted below.				

This Receipt Is Valid Only When Signed by an Officer of the Company.

Advances have been made and liability incurred on such goods, as follows:

____

____

The property covered by this receipt has NOT been insured by this company for the benefit of the depositor against fire or any other casualty.

(This clause to be omitted from forms used in those states where warehousemen are required by law to insure goods.)

American Warehouse Company claims a lien for all lawful charges for storage and preservation of the goods; also for all lawful claims for money advanced, interest, insurance, transportation, labor, weighing, coopering and other charges and expenses in relation to such goods.

AMERICAN WAREHOUSE COMPANY

By__________

THE GOODS MENTIONED BELOW ARE HEREBY RELEASED FROM THIS RECEIPT FOR DELIVERY FROM WAREHOUSE. ANY UNRELEASED BALANCE OF THE GOODS IS SUBJECT TO A LIEN FOR UNPAID CHARGES AND ADVANCES ON THE RELEASED PORTION.

DELIVERIES

DATE	LOT NUMBER	QUANTITY RELEASED		SIGNATURE	QUANTITY DUE ON RECEIPT

merchandise is necessary upon receipt at the warehouse.

Packaging—There are two classes of packaging service performed by a warehouseman. First, what is known as unavoidable packaging caused by damage to the original container. Second, the packaging required as the result of changes such as delivery to the customer in larger or smaller units than the original packages, shipments for export, etc.

AMERICAN WAREHOUSE COMPANY
STREET ADDRESS • CITY & AMERICA 00000
TELEPHONE: (312) – 123-4567

ORIGINAL
NON-NEGOTIABLE WAREHOUSE RECEIPT

AMERICAN WAREHOUSE COMPANY claims a lien for all lawful charges for storage and preservation of the goods; also for all lawful claims for money advanced, interest, insurance, transportation, labor, weighing, coopering and other charges and expenses in relation to such goods, and for the balance on any other accounts that may be due. The property covered by this receipt has NOT been insured by this Company for the benefit of the depositor against fire or any other casualty.

DOCUMENT NUMBER
DATE
CUSTOMER NUMBER
CUSTOMER ORDER NO.
WAREHOUSE NO.

RECEIVED FROM

THIS IS TO CERTIFY THAT WE HAVE RECEIVED the goods listed hereon in apparent good order, except as noted herein (contents, condition and quality unknown), SUBJECT TO ALL TERMS AND CONDITIONS INCLUDING LIMITATION OF LIABILITY HEREIN AND ON THE REVERSE HEREOF. Such property to be delivered to THE DEPOSITOR upon the payment of all storage, handling and other charges. Advances have been made and liability incurred on these goods as follows:

FOR ACCOUNT OF

DELIVERING CARRIER	CARRIER NUMBER	PREPAID/COLLECT	SHIPPERS NUMBER

QUANTITY	SAID TO BE OR CONTAIN (CUSTOMER ITEM NO., WAREHOUSE ITEM NO., LOT NUMBER, DESCRIPTION, ETC.)	WEIGHT	RATE CODE	STORAGE RATE / HANDLING RATE	DAMAGE & EXCEPTIONS
	TOTALS				

AWA Seal of Security

NO DELIVERY WILL BE MADE ON THIS RECEIPT EXCEPT ON WRITTEN ORDER.

AMERICAN WAREHOUSE COMPANY
BY
AUTHORIZED SIGNATURE

Agent of the Shipper—Acting as the agent of the shipper by preparing bills of lading, paying freight charges, invoicing customers and performing many other clerical services.

Facilities of Merchandise Warehouses

From a transportation standpoint most of the modern warehouses have private railway sidings for the accommodation of from one to several hundred cars depending on the size of the warehouse and the city in which it is located. In addition to the rail facilities most of the warehouses have truck terminal facilities and many of the warehouses located at the ports have docking facilities in addition to the rail and truck facilities. Such transportation facilities make it possible to load and unload cars, trucks and boats promptly and economically because of minimum labor costs and transportation charges.

The modern warehouse is a single-level building of fire resistant construction. It is equipped with sprinkler systems, automatic burglar and fire alarms and watchman service. All of this precaution reduces the fire and theft hazard and results in lower insurance rates. In addition to thousands of square feet of open floor space for storage use, many warehouses have vaults for valuables, heated rooms, cooler rooms, fumigation and reconditioning facilities, and branch office and showroom facilities.

The modern warehouse is equipped with many labor-saving devices, i.e., latest type scales, portable conveyors, traveling overhead electric cranes, electric lift trucks with skid platforms, and numerous other handling devices. The use of this equipment results in economy in time and labor when handling merchandise from the car to storage space, in storing and piling and in distribution. Special material handling capabilities can be implemented when large scale operations are arranged.

Many warehouses own and operate their own fleet of trucks, thus insuring their customers a flexible and rapid cartage and store-door delivery service. In most instances, a twenty-four hour delivery to jobber, department, chain or drug stores and consumers is systematically maintained within a certain radius of the warehouse.

In addition to the physical set-up described in the foregoing, the modern warehouse maintains a competent office and clerical personnel to give its customers a complete service on their storage and distribution requirements. This service for example, includes: the handling of client's mail and telephone calls promptly; maintaining a complete stock control;

a filing system which gives a full record of each transaction; monthly inventories; and the use of its traffic and legal departments. Some warehouses have electronic data processing equipment and their service may include a direct access by the depositor's computer.

Theory of Warehouse Charges

Nowhere is the function of service more significant than in the field of storage, warehousing and distribution. The exacting requirements of these services call for the utmost efficiency, economy and complete, modern facilities in every department. Just as the benefits to be derived from efficient and complete warehousing are of primary importance to tradesmen, so the warehousemen are vitally interested in obtaining the largest possible revenue for their services.

Warehousing charges must be just and reasonable, not alone to the depositor but to the warehousemen as well, in order that they may derive an adequate return on their investment. This involves consideration of a number of factors in establishing warehouse rates or charges.

Basis of Storage and Handling Rates—In the early days of the warehouse industry, storage and handling rates were the result of shrewd guesses which sometimes resulted in discriminatory practices. With the growth of the merchandise warehouse business the warehousemen's association, based upon experience, developed "a standardized basis for rates form which the individual warehouseman could compute his rates."

Three important factors which enter into the making of a storage rate are:

1. The value of the goods, because of the risk involved.
2. Nature of the merchandise—the size and shape, adaptability for tiering, its danger of contaminating or being contaminated.
3. Requirements of the merchandise as to cold and heat, dry or moist air, etc.

In addition to these important factors the "standardized basis" included the factors of: floor space, floor load, density and shape of packages and height of pile. Based upon these factors and various rates per square foot of floor space, the determination of earnings per square foot of floor space having been left to the individual warehouseman, a series of "rate and classification tables" were established. These tables combined the known facts of any particular commodity and contained an equitable relation between different commodities, so that each commodity or class of commodities, whether stored in bulk or in packages would bear its proper portion of the cost of service and yield its proper net revenue.

The rate on a given lot of merchandise was determined by the use of the unit known as the "standard warehouse pile." This unit, also known as the "lot unit," was the "greatest quantity of the commodity that could be practically stored, in bulk, on 120 square feet of floor space, at not exceeding 250 pounds weight per square foot, or piled not to exceed seven feet-three inches high." From this unit, set-up for each commodity, and multiplied by the "rate per foot of space" for the particular warehouse, was obtained the amount to be charged for this storage. This rate would be the charge for the ideal package—that is the package which in piles seven feet-three inches high would weigh 250 pounds per square foot. If a package weighed less than the standard per cubic foot, the rate was based on its cubic measurement, and if it weighed more than the standard of density, the rate was based on its actual weight. Generally the unit which yields the greater revenue is charged.

The factors used to determine the handling rate were: (1) volume; (2) variety of packages; (3) type of packaged, that is, style and shape; (4) density; and (5) weight.

It should be remembered that the "standardized basis" was developed a number of years ago at a time when most warehouses were multi-level buildings with elevators and a very limited ceiling height and floor load. Although it is still used in some cases today the cost factors that were applicable then have changed drastically.

Merchandise Warehouse Rates and Charges

The rates charged by public warehouses are established independently by each warehouseman. Unlike the freight rates of common carriers the rates for storage and handling in public warehouses are not standard. In almost all cases the rates are the result of negotiations between the warehouseman and the depositor. In actual practice a manufacturer desiring to store merchandise in a particular public warehouse will provide the warehouseman with a complete description of the goods to be stored and the facilities and services required and request a rate quotation. The warehouseman's quotation will usually be furnished on a form such as shown on page 000. This form contains the Standard Contract Terms and Conditions and when accepted and signed becomes the contract between the parties.

Storage rates are usually expressed per package or piece or some other unit of size or weight per month. Handling charges, which cover only the labor involved in receiving the goods, placing them in storage and returning them to the shipping platform, may be expressed in the same manner.

Bills for charges are usually rendered upon receipt of the goods and on a monthly basis thereafter. In most cases the warehouse will show the charge for each service performed as a separate item on their bill.

Warehouseman's Liability

Warehousemen's liability has its foundations in the law of bailments. Bailment is defined in Cyclopedic Law Dictionary as:

> "A delivery of something of a personal nature by one party to another, to be held according to the purpose or object of the delivery, and to be returned or delivered over when that purpose is accomplished."

A bailment relationship arises whenever title to personal property remains in one party (the bailor) and possession of the property is given to another party (the bailee). When we speak of the liability of a warehouseman we are in effect, speaking of the liability of a bailee for hire. The degree of liability is identical. A warehouseman, or bailee for hire, has a certain amount of responsibility for the safekeeping of goods in his possession. He is bound to use reasonable care in the protection of the goods, and he is liable for any loss or damage that results from his fault or negligence.

Section 7-204 of the Uniform Commercial Code (reproduced at the end of this chapter) defines the liability and the limitation of damages of warehousemen and its provisions are incorporated in Section 11 of the Standard Contract Terms and Conditions. This section should be read very carefully, particularly paragraph (C). This paragraph permits the warehouseman to specify in the contract the limit of his liability. This is usually expressed as a specific amount per article, per piece, per item or unit of weight. In addition the warehouseman will indicate the charge for additional coverage should the depositor desire it.

Claims for loss or injury to the goods must be filed in writing with the warehouse within 60 days after delivery of the goods by the warehouse. The filing of claim is a condition precedent to suit and all actions must be begun within nine months after delivery.

Insurance

The rates charged by the warehouse do not include fire or other insurance. The merchandise in the warehouse is not insured by the warehousemen unless the storer given written order to do so, stating the amount and kind of insurance desired.

Insurance coverage for warehouses includes fire insurance as a matter of course, the other hazards, i.e., floods, tornado, earthquake, theft and burglary, sprinkler-leakage and numerous others, are included in the insurance coverage where such liability exists, the stipulations, descriptions and conditions of the fire insurance policy being the basis for underwriting policies for all these hazards.

Fire-proof construction, sprinkler system, watchmen service, etc. all tend to reduce the fire hazards and result in favorable insurance rates for the warehouse and the merchandise stored therein. Warehouses with low insurance ratings indicate lessened hazards for the merchandise in store. The cost of insurance on warehoused merchandise is very important and the insurance rating should be a guide to the selection of a warehouse so as to obtain favorable insurance rates and lessened hazards for the merchandise to be stored.

The insurance rate on the warehouse is the base for the insurance rate on the commodity stored. To this base rate is added, for each commodity stored, its own rate, which is known as its individual "susceptibility charge." This charge may be obtained from the insurance underwriters.

Railroad Warehouse Practices

The practice of the railroads, by direct or indirect ownership of warehouses and other devices, performing warehouse services in competition with private commercial warehouses has been brought to the attention of the Interstate Commerce Commission in a number of cases because of alleged discriminatory practices.

In *McCormick Warehouse Co. v. P.R.R. Co.,* (95 I.C.C. 301), it was alleged that the railroad employed a certain warehouse company, one-third of the capital stock of which was controlled by the carrier, to perform terminal services in connection with the loading and unloading of carload package freight received in pool cars at Baltimore, Md., and refused to employ the McCormick Warehouse Company to perform similar services, for which the railroad made an allowance. The Commission found that no unjust discrimination or undue prejudice was shown to exist. However, upon rehearing, 148 I.C.C. 299, it issued an order, instructing the railroad to discontinue its practice, as the allowances to the railroad controlled warehouse were nothing more than a device to attempt to lend legality to the payment of rebates. A bill to enjoin this order was filed in the United States district court by the warehouse used by the carrier, but in *Terminal Warehouse Co. v. United States,* (31 F. (2d) 951), the order was upheld and said bill dismissed. After complying with the order the case was reopened, upon petition of the carrier, to determine whether said order should be modified with respect to allowances to the warehouse for loading and unloading water-borne freight at Baltimore. In the reopened case, 191 I.C.C. 727, the Commission found that modification of its previous order was not warranted.

In *James Gallagher et al. v. P.R.R. Co.,* (160 I.C.C. 563), a situation at Philadelphia, Pa., similar to the McCormick situation at Baltimore, the Commission ordered the carrier (1) to cease and desist from publishing or making such allowances, and (2) to cancel tariff provisions which make the warehouses used by them a part of their respective station facilities at Philadelphia. This order of the Commission was considered by the Supreme Court in *Merchants Warehouse Co. et al. v. United States, Interstate Commerce Commission et al.,* (283 U.S. 501), and the validity of the Commission's order upheld. The court also held that this case was one in which the carriers should not be left free to remove the discrimination by extending the benefits of the allowances to the competing commercial warehouses, since:

> "the allowances are for services which include the assembling and distribution of carloads from or into less than carload lots, the objections to them would not be removed by extending them to some additional shippers of carloads at carload rates, but not to all, nor even to all under existing tariffs and classifications for the handling of carloads. As the Commission found that appellants' warehouses are not public freight stations in fact but are such in name only, it rightly secured the discontinuance of the discrimination by ordering the carriers to cease employing the means by which it had been accomplished."

In an investigation on its own motion, *Ex Parte No. 104, Part VI, Warehousing and Storage of Property by Carriers at Port of New York, N.Y.,* (198 I.C.C. 134), the Commission held that: (1) Carriers' warehousing and storage practices, charges assessed and allowances made in connection therewith at the Port of New York district, found to dissipate their funds and revenues, not to be in conformity with efficient and economical management as contemplated by the Interstate Commerce Act, and not in the public interest. Certain of such practices, charges, and allowances of the individual carriers, cited in the report, found to result in violations of the Interstate Commerce Act and afford reasonable ground for the belief that the Elkins Act is violated. Carriers admonished to take prompt corrective action. (2) Carriers serving other ports and terminals admonished to adjust their practices and charges in conformity with the principles herein announced. This order of the Commission was sustained by the Supreme Court in *Baltimore & Ohio R.R. Co. v. United States,* (305 U.S. 507), wherein the court said:

> ". . . if the service is non-transportation, the fact that it is in a tariff does not save it from the condemnation of Section 6(7). That section forbids receiving a less compensation for transportation than the tariff. The loss on in-transit warehousing, entered into to secure the rail-haul, results in lowered receipts for the transportation and in violation of the section. Some shippers are not in a position to avail themselves of the below-cost in-transit service. They must pay the full transportation rate, without any offset from the warehousing. This discrimination between shippers is unlawful and the remedy applied by the order valid in these circumstances. . . ."

The foregoing cases are sufficient to illustrate the attitude of the Commission and the Courts as to railroad warehousing in competition with private warehousing. In its conclusions in the New York Warehouse case, *supra,* at pages 194 and 195 of the report, the Commission clearly defines railroad storage under the Act as follows:

> "The term 'transportation' as used in section 1(3) of the Act includes the receipt, delivery, elevation and transfer in transit, ventilation, refrigeration or icing, storage, and handling of property to provide and furnish such transportation upon reasonable request therefor. While storage of property is clearly within the transportation service which carriers are obligated to furnish, their duty under these provisions extends only to that storage which is necessarily incidental to transporting such property. To be incidental business, the storage must be preliminary either to immediate transportation or immediate removal."

Transportation Considerations

There is considerably more to this subject of warehousing than the mere selection of a particular warehouse based upon the services provided and the rates applicable. Transportation plays an extremely important role in the selection of both the location of the warehouse and the particular warehouse to be used.

One of the first considerations would probably be the rates to and from the warehouse. In other words the carload or truckload rates into the warehouse point and the LCL or LTL rates from the warehouse location to the ultimate destination. The aggregate of these charges plus the cost of warehousing should be less than the through LCL or LTL charges from point of manufacture to ultimate destination, except in those cases where service or availability of product is the prime consideration.

Another important location factor is the specific location of the warehouse within the city. A location outside the switching limits or the terminal area could result in additional costs.

Another factor related to location would be the transportation services available. If rail service is the dominant factor on the inbound move and the warehouse has a private siding served by one of the railroads a question to be resolved would be, is there a reciprocal switching arrangement which would permit routing via all railroads serving the city without incurring additional switching costs. Where both the inbound and outbound movements are handled by motor carrier it would be important to determine the extent of the loading and unloading facilities at the warehouse so as to avoid serious delays and the possibility of detention charges due to congestion. The number of motor carriers serving the city and the availability of equipment for outbound movements is also worthy of consideration.

With regard to freight rates applicable to and from the warehouse questions occasionally arise where one of the movements is wholly within the borders of one state. If both the interstate and intrastate rates are the same, there is no problem; however where there are two scales of rates there could be a question as to which scale applies. For example, assume that carload shipments are forwarded from New York to Chicago, Ill., and placed in a warehouse at that point for ultimate distribution to points in Illinois and surrounding states. Are the shipments from Chicago to points in Illinois intrastate or, by virtue of the original move from New York, interstate? This question has been before the Commission and the courts many, many times and it would seem from the multitude of cases that every conceivable question should have been satisfactorily answered but such is not the case. Where this situation is encountered the question should be resolved before shipments move so as to avoid the possibility of balance due bills being presented at some later date.

APPENDIX TO CHAPTER 11

STANDARD CONTRACT TERMS AND CONDITIONS FOR MERCHANDISE WAREHOUSEMEN

Approved and promulgated by the American Warehousemen's Association
October, 1968

ACCEPTANCE—Sec. 1

(a) This contract and rate quotation including accessorial charges endorsed on or attached hereto must be accepted within 30 days from the proposal date by signature of depositor on the reverse side of the contract. In the absence of written acceptance, the act of tendering goods described herein for storage or other services by warehouseman within 30 days from the proposal date shall constitute such acceptance by depositor.

(b) In the event that goods tendered for storage or other services do not conform to the description contained herein, or conforming goods are tendered after 30 days from the proposal date without prior written acceptance by depositor as provided in paragraph (a) of this section, warehouseman may refuse to accept such goods. If warehouseman accepts such goods, depositor agrees to rates and charges as may be assigned and invoiced by warehouseman and to all terms of this contract.

(c) This contract may be cancelled by either party upon 30 days written notice and is cancelled if no storage or other services are performed under this contract for a period of 180 days.

SHIPPING—Sec. 2

Depositor agrees not to ship goods to warehouseman as the named consignee. If, in violation of this agreement, goods are shipped to warehouseman as named consignee, depositor agrees to notify carrier in writing prior to such shipment, with copy of such notice to the warehouseman, that warehouseman named to consignee is a warehouseman and has no beneficial title or interest in such property and depositor further agrees to indemnify and hold harmless warehouseman from any and all claims for unpaid transportation charges, including undercharges, demurrage, detention or charges of any nature, in connection with goods so shipped. Depositor further agrees that, if it fails to notify carrier as required by the next preceding sentence, warehouseman shall have the right to refuse such goods and shall not be liable or responsible for any loss, injury or damage of any nature to, or related to, such goods. Depositor agrees that all promises contained in this section will be binding on depositor's heirs, successors and assigns.

TENDER FOR STORAGE—Sec. 3

All goods for storage shall be delivered at the warehouse properly marked and packaged for handling. The depositor shall furnish, at or prior to such delivery, a manifest showing marks, brands, or sizes to be kept and accounted for separately, and the class of storage and other services desired.

STORAGE PERIOD AND CHARGES—Sec. 4

(a) All charges for storage are per package or other agreed unit per month.

(b) Storage charges become applicable upon the date that warehouseman accepts care, custody and control of the goods, regardless of unloading date or date of issue of warehouse receipt.

(c) Except as provided in paragraph (d) of this section, a full month's storage charge will apply on all goods received between the first and the 15th, inclusive, of a calendar month; one-half month's storage charge will apply on all goods received between the 16th and the last day, inclusive, of a calendar month, and a full month's storage charge will apply to all goods in storage on the first day of the next and succeeding calendar months. All storage charges are due and payable on the first day of storage for the initial month and thereafter on the first day of the calendar month.

(d) When mutually agreed by the warehouseman and the depositor, a storage month shall extend from a date in one calendar month to, but no including, the same date of the next and all succeeding months. All storage charges are due and payable on the first day of the storage month.

TRANSFER, TERMINATION OF STORAGE, REMOVAL OF GOODS—Sec. 5

(a) Instructions to transfer goods on the books of the warehouseman are not effective until delivered to and accepted by warehouseman, and all charges up to the time transfer is made are chargeable to the depositor of record. If a transfer involves rehandling the goods, such will be subject to a charge. When goods in storage are transferred from one party to another through issuance of a new warehouse receipt, a new storage date is established on the date of transfer.

(b) The warehouseman reserves the right to move, at his expense, 14 days after notice is sent by certified or registered mail to the depositor of record or to the last known holder of the negotiable warehouse receipt, any goods in storage from the warehouse in which they may be stored to any other of his warehouses; but if such depositor or holder takes delivery of his goods in lieu of transfer, no storage charge shall be made for the current storage month. The warehouseman may, without notice, move goods within the warehouse in which they are stored.

(c) The warehouseman may, upon written notice to the depositor of record and any other person known by the warehouseman to claim an interest in the goods, require the removal of any goods by the end of the next succeeding storage month. Such notice shall be given to the last known place of business or abode of the person to be notified. If goods are not removed before the end of the next succeeding storage month, the warehouseman may sell them in accordance with applicable law.

(d) If the warehouseman in good faith believes that the goods are about to deteriorate or decline in value to less than the amount of warehouseman's lien before the end of the next succeeding storage month, the warehouseman may specify in the notification any reasonable shorter time for removal of the goods and in case the goods are not removed, may sell them at public sale held one week after a single advertisement or posting as provided by law.

(e) If as a result of a quality or condition of the goods of which the warehouseman had no notice at the time of deposit the goods are a hazard to other property or to the warehouse or to persons, the warehouseman may sell the goods at public or private sale without advertisement on reasonable notification to all persons known to claim an interest in the goods. If the warehouseman after a reasonable effort is unable to sell the goods he may dispose of them in any lawful manner and shall incur no liability by reason of such disposition. Pending such disposition, sale or return of the goods, the warehouseman may remove the goods from the warehouse and shall incur no liability by reason of such removal.

HANDLING—Sec. 6

(a) The handling charge covers the ordinary labor involved in receiving goods at warehouse door, placing goods in storage, and returning goods to warehouse door. Handling charges are due and payable on receipt of goods.

(b) Unless otherwise agreed, labor for unloading and loading goods will be subject to a charge. Additional expenses incurred by the warehouseman in receiving and handling damaged goods, and an additional expense in unloading from or loading into cars or other vehiches not at warehouse door will be charged to the depositor.

(c) Labor and materials used in loading rail cars or other vehicles are chargeable to the depositor.

(d) When goods are ordered out in quantities less than in which received, the warehouseman may make an additional charge for each order or each item of an order.

(e) The warehouseman shall not be liable for demurrage, delays in unloading inbound cars, or delays in obtaining and loading cars for outbound shipment unless warehouseman has failed to exercise reasonable care.

DELIVERY REQUIREMENTS—Sec. 7

(a) No goods shall be delivered or transferred except upon receipt by the warehouseman of complete instructions properly signed by the depositor. However, when no negotiable receipt is outstanding, goods may be delivered upon instructions by telephone in accordance with a prior written authorization, but the warehouseman shall not be responsible for loss or error occasioned thereby.

(b) When a negotiable receipt has been issued no goods covered by that receipt shall be delivered, or transferred on the books of the warehouseman, unless the receipt, properly endorsed, is surrendered for cancellation, or for endorsement of partial delivery thereon. If a negotiable receipt is lost or destroyed, delivery of goods may be made only upon order of a court of competent jurisdiction and the posting of security approved by the court as provided by law.

(c) When goods are ordered out a reasonable time shall be given the warehouseman to carry out instructions, and if he is unable because of acts of God, war, public enemies, seizure under legal process, strikes, lockouts, riots and civil commotions, or any reason beyond the warehouseman's control, or because of loss or destruction of goods for which warehouseman is not liable, or because of any other excuse provided by law, the warehouseman shall not be liable for failure to carry out such instructions and goods remaining in storage will continue to be subject to regular storage charges.

EXTRA SERVICES (Special Services)—Sec. 8

(a) Warehouse labor required for services other than ordinary handling and storage will be charged to the depositor.

(b) Special services requested by depositor including but not limited to compiling of special stock statements; reporting marked weights, serial numbers or other data from packages; physical check of goods; and handling transit billing will be subject to a charge.

(c) Dunnage, bracing, packing materials or other special supplies may be provided for the depositor at a charge in addition to the warehouseman's cost.

(d) By prior arrangement, goods may be received or delivered during other than usual business hours, subject to a charge.

(e) Communication expense including postage, teletype, telegram, or telephone, will be charged to the depositor if such concern more than normal inventory reporting or if, at the request of the depositor, communications are made by other than regular United States Mail.

BONDED STORAGE—Sec. 9

(a) A charge in addition to regular rates will be made for merchandise in bond.

(b) Where a warehouse receipt covers goods in U.S. Customs bond, such receipt shall be void upon the termination of the storage period fixed by law.

MINIMUM CHARGES—Sec. 10

(a) A minimum handling charge per lot and a minimum storage charge per lot per month will be made. When a warehouse receipt covers more than one lot or when a lot is in assortment, a minimum charge per mark, brand, or variety will be made.

(b) A minimum monthly charge to one account for storage and/or handling will be made. This charge will apply also to each account when one customer has several accounts, each requiring separate records and billing.

LIABILITY AND LIMITATION OF DAMAGES—Sec. 11

(a) The warehouseman shall not be liable for any loss or injury to goods stored however caused unless such loss or injury resulted from the failure by the warehouseman to exercise such care in regard to them as a reasonably careful man would exercise under like circumstances and warehouseman is not liable for damages which could not have been avoided by the exercise of such care.

(b) Goods are not insured by warehouseman against loss or injury however caused.

(c) The depositor declares that damages are limited to ____________________, provided, however, that such liability may at the time of acceptance of this contract as provided in Section 1 be increased on part or all of the goods hereunder in which event a monthly charge of ____________________ will be made in addition to the regular monthly storage charge.

NOTICE OF CLAIM AND FILING OF SUIT—Sec. 12

(a) Claims by the depositor and all other persons must be presented in writing to the warehouseman within a reasonable time, and in no event longer than either 60 days after delivery of the goods by the warehouseman or 60 days after depositor of record or the last known holder of a negotiable warehouse receipt is notified by the warehouseman that loss or injury to part or all of the goods has occurred, whichever time is shorter.

(b) No action may be maintained by the depositor or others against the warehouseman for loss or injury to the goods stored unless timely written claim has been given as provided in paragraph (a) of this section and unless such action is commenced either within nine months after date of delivery by warehouseman or within nine months after depositor of record or the last known holder of a negotiable warehouse receipt is notified that loss or injury to part or all of the goods has occurred, whichever time is shorter.

(c) When goods have not been delivered, notice may be given of known loss or injury to the goods by mailing of a registered or certified letter to the depositor of record or to the last known holder of a negotiable warehouse receipt. Time limitations for presentation of claim in writing and maintaining of action after notice begin on the date of mailing of such notice by warehouseman."

Uniform Commercial Code
Article 7
Warehouse Receipts, Bills of Lading and Other Documents of Title

Part I
General

Sec. 7-101. Short Title.

This Article shall be known and may be cited as Uniform Commercial Code—Documents of Title.

Sec. 7-102. Definitions and Index of Definitions.

(1) In this Article, unless the context otherwise requires:

(a) "Bailee" means the person who by a warehouse receipt, bill of lading, or other document of title acknowledges possession of goods and contracts to deliver them.

(b) "Consignee" means the person named in a bill to whom or to whose order the bill promises delivery.

(c) "Consignor" means the person named in a bill as the person from whom the goods have been received for shipment.

(d) "Delivery order" means a written order to deliver goods directed to a warehouseman, carrier or other person who in the ordinary course of business issues warehouse receipts or bills of lading.

(e) "Document" means document of titles as defined in the general definitions in Article 1 (Section 1-201).

(f) "Goods" means all things which are treated as movable for the purposes of a contract of storage or transportation.

(g) "Issuer" means a bailee who issues a document except that in relation to an unaccepted delivery order it means the person who orders the possessor of goods to deliver. Issuer includes any person for whom an agent or employee purports to act in issuing a document if the agent or employee has real or apparent authority to issue documents, notwithstanding that the issuer received no goods or that the goods were misdescribed or that in any other respect the agent or employee violated his instructions.

(h) "Warehouseman" is a person engaged in the business of storing goods for hire.

(2) Other definitions applying to this Article or to specified Parts thereof, and the sections in which they appear are:

"Duly negotiate." Section 7-501.
"Person entitled under the document." Section 7-403(4).

(3) Definitions in other Articles applying to this Article and the sections in which they appear are:

"Contract for sale." Section 2-106.
"Overseas." Section 2-323.
"Receipt" of goods. Section 2-103.

(4) In addition Article 1 contains general definitions and principles of construction and interpretation applicable throughout this Article.

Sec. 7-103. Relation of Article to Treaty, Statute, Tariff, Classification or Regulation.

To the extent that any treaty or statute of the United States, regulatory statute of this State or tariff, classification or regulation filed or issued pursuant thereto is applicable, the provisions of this article are subject thereto.

Sec. 7-104. Negotiable and Non-Negotiable Warehouse Receipt, Bill of Lading or other Document of Title.

(1) A warehouse receipt, bill of lading or other document of title is negotiable

(a) if by its terms the goods are to be delivered to bearer or to the order of a named person; or

(b) where recognized in overseas trade, if it runs to a named person or assigns.

(2) Any other document is non-negotiable. A bill of lading in which it is stated that the goods are consigned to a named person is not made negotiable by a provision that the goods are to be delivered only against a written order signed by the same or another person.

Sec. 7-105. Construction Against Negative Implication.

The omission from either Part 2 or Part 3 of this Article of a provision corresponding to a provision made in the other Part does not imply that a corresponding rule of law is not applicable.

Part 2
Warehouse Receipts: Special Provisions

Sec. 7-201. Who May Issue a Warehouse Receipt; Storage Under Government Bond.

(1) A warehouse receipt may be issued by any warehouseman.

(2) Where goods including distilled spirits and agricultural commodities are stored under a statute requiring a bond against withdrawal or a license for the issuance of receipts in the nature of warehouse receipts, a receipt issued for the goods has like effect as a warehouse receipt even though issued by a person who is the owner of the goods and is not a warehouseman.

Sec. 7-202. Form of Warehouse Receipt; Essential Terms; Optional Terms.

(1) A warehouse receipt need not be in any particular form.

(2) Unless a warehouse receipt embodies within its written or printed terms each of the following, the warehouseman is liable for damages caused by the omission to a person injured thereby:

(a) the location of the warehouse where the goods are stored;

(b) the date of issue of the receipt;

(c) the consecutive number of the receipt;

(d) a statement whether the goods received will be delivered to the bearer, to a specified person, or to a specified person or his order;

(e) the rate of storage and handling charges, except that where goods are stored under a field warehousing arrangement a statement of that fact is sufficient on a non-negotiable receipt;

(f) a description of the goods or of the packages containing them;

(g) the signature of the warehouseman, which may be made by his authorized agent;

(h) if the receipt is issued for goods of which the warehouseman is owner, either solely or jointly or in common with others, the fact of such ownership; and

(i) a statement of the amount of advances made and of liabilities incurred for which the warehouseman claims a lien or security interest (Section 7-209). If the precise amount of such advances made or of such liabilities incurred is, at the time of the issue of the receipt, unknown to the warehouseman or to his agent who issues it, a statement of the fact that advances have been made or liabilities incurred and the purpose thereof is sufficient.

(3) A warehouseman may insert in his receipt any other terms which are not contrary to the provisions of this Act and do not impair his obligations of delivery (Section 7-403) or his duty of care (Section 7-204). Any contrary provisions shall be ineffective.

Sec. 7-203. Liability for Non-Receipt or Misdescription.

A party to or purchaser for value in good faith of a document of title other than a bill of lading relying in either case upon the description therein of the goods may recover from the issuer damages caused by the non-receipt or misdescription of the goods, except to the extent that the document conspicuously indicates that the issuer does not know whether any part or all of the goods in fact were received or conform to the description, as where the description is in terms of marks or labels or kind, quantity or condition, or the receipt or description is qualified by "contents, condition and quality unknown," "said to contain" or the like, if such indication be true, or the party or purchaser otherwise has notice.

Sec. 7-204. Duty of Care; Contractual Limitation of Warehouseman's Liability.

(1) A warehouseman is liable for damages for loss of or injury to the goods caused by his failure to exercise such care in regard to them as a reasonably careful man would exercise under like circumstances, but unless otherwise agreed, he is not liable for damages which could not have been avoided by the exercise of such care.

(2) Damages may be limited by a term in the warehouse receipt or storage agreement limiting the amount of liability in case of loss or damage, and setting forth a specific liability per article or item, or value per unit of weight, beyond which the warehouseman shall not be liable; provided, however, that such liability may on written request of the bailor at the time of signing such storage agreement or within a reasonable time after receipt of the warehouse receipt be increased on part or all of the goods thereunder, in which event increased rates may be charged based on such increased valuation, but that no such increase shall be permitted contrary to a lawful limitation of liability contained in the warehouseman's tariff, if any. No such limitation is effective with respect to the warehouseman's liability for conversion to his own use.

(3) Reasonable provisions as to the time and manner of presenting claims and instituting actions based on the bailment may be included in the warehouse receipt or tariff.

(4) This section does not impair or repeal . . .

Note: Insert in subsection (4) a reference to any statute which imposes a higher responsibility upon the warehouseman or invalidates contractual limitations which would be permissible under this Article.

Sec. 7-205. Title Under Warehouse Receipt Defeated in Certain Cases.

A buyer in the ordinary course of business of fungible goods sold and delivered by a warehouseman who is also in the business of buying and selling such goods takes free of any claim under a warehouse receipt even though it has been duly negotiated.

Sec. 7-206. Termination of Storage at Warehouseman's Option.

(1) A warehouseman may on notifying the person on whose account the goods are held and any other person known to claim an interest in the goods require payment of any charges and removal of the goods from the warehouse at the termination of the period of storage fixed by the document, or, if no period is fixed, within a stated period not less than thirty days after the notification. If the goods are not removed before the date specified in the notification, the warehouseman may sell them in accordance with the provisions of the section on enforcement of a warehouseman's lien (Section 7-210).

(2) If a warehouseman in good faith believes that the goods are about to deteriorate or decline in value to less than the amount of his lien within the time prescribed in subsection (1) for notification, advertisement and sale, the warehouseman may specify in the notification any reasonable shorter time for removal of the goods and in case the goods are not removed, may sell them at public sale held not less than one week after a single advertisement or posting.

(3) If as a result of a quality or condition of the goods of which the warehouseman had no notice at the time of deposit the goods are a hazard to other property or to the warehouse or to persons, the warehouseman may sell the goods at public or private sale without advertisement on reasonable notification to all persons known to claim an interest in the goods. If the warehouseman after a reasonable effort is unable to sell the goods he may dispose of them in any lawful manner and shall incur no liability by reason of such disposition.

(4) The warehouseman must deliver the goods to any person entitled to them under this Article upon due demand made at any time prior to sale or other disposition under this section.

(5) The warehouseman may satisfy his lien from the proceeds of any sale or disposition under this section but must hold the balance for delivery on the demand of any person to whom he would have been bound to deliver the goods.

Sec. 7-207. Goods Must Be Kept Separate; Fungible Goods.

(1) Unless the warehouse receipt otherwise provides, a warehouseman must keep separate the goods covered by each receipt so as to permit at all times identification and delivery of those goods except that different lots of fungible goods may be commingled.

(2) Fungible goods so commingled are owned in common by the persons entitled thereto and the warehouseman is severally liable to each owner for that owner's share. Where because of overissue a mass of fungible goods is insufficient to meet all the receipts which the warehouseman has issued against it, the persons entitled include all holders to whom overissued receipts have been duly negotiated.

Sec. 7-208. Altered Warehouse Receipts.

Where a blank in a negotiable warehouse receipt has been filled in without authority, a purchaser for value and without notice of the want of authority may treat the insertion as authorized. Any other unauthorized alteration leaves any receipt enforceable against the issuer according to its original tenor.

Sec. 7-209. Lien of Warehouseman.

(1) A warehouseman has a lien against the bailor on the goods covered by a warehouse receipt or on the proceeds thereof in his possession for charges for storage or transportation (including demurrage and terminal charges), insurance, labor, or charges present or future in relation to the goods, and for expenses necessary for preservation of the goods or reasonably incurred in their sale pursuant to law. If the person on whose account the goods are held is liable for like charges or expenses in relation to other goods whenever deposited and it is stated in the receipt that a lien is claimed for charges and expenses in relation to other goods, the warehouseman also has a lien against him for such charges and expenses whether or not the other goods have been delivered by the warehouseman. But against a person to whom a negotiable warehouse receipt is duly negotiated a warehouseman's lien is limited to charges in an amount or at a rate specified on the receipt or if no charges are so specified then to a reasonable charge for storage of the goods covered by the receipt subsequent to the date of the receipt.

(2) The warehouseman may also reserve a security interest against the bailor for a maximum amount specified on the receipt for charges other than those specified in subsection (1), such as for money advanced and interest. Such a security interest is governed by the Article on Secured Transactions (Article 9).

(3) A warehouseman's lien for charges and expenses under subsection (1) or a security interest under subsection (2) is also effective against any person who so entrusted the bailor with possession of the goods that a pledge of them by him to a good faith purchaser for value would have been valid but is not effective against a person to whom the document confers no right in the goods covered by it under Section 7-503.

(4) A warehouseman loses his lien on any goods which he voluntarily delivers or which he unjustifiably refuses to deliver.

Sec. 7-210. Enforcement of Warehouseman's Line.

(1) Except as provided in subsection (2), a warehouseman's lien may be enforced by public or private sale of the goods in bloc or in parcels, at any time or place and on any terms which are commercially reasonable, after notifying all persons known to claim an interest in the goods. Such notification

must include a statement of the amount due, the nature of the proposed sale and the time and place of any public sale. The fact that a better price could have been obtained by a sale at a different time or in a different method from that selected by the warehouseman is not of itself sufficient to establish that the sale was not made in commercially reasonable manner. If the warehouseman either sells the goods in the usual manner in any recognized market therefore, or if he sells at the price current in such market at the time of his sale, or if he has otherwise sold in conformity with commercially reasonable practices among dealers in the type of goods sold, he has sold in a commercially reasonable manner. A sale of more goods than apparently necessary to be offered to insure satisfaction of the obligations is not commercially reasonable except in cases covered by the preceding sentence.

(2) A warehouseman's lien on goods other than goods stored by a merchant in the course of his business may be enforced only as follows:

(a) All persons known to claim an interest in the goods must be notified.

(b) The notification must be delivered in person or sent by registered letter to the last known address of any person to be notified.

(c) The notification must include an itemized statement of the claim, a description of the goods subject to the lien, a demand for payment within a specified time not less than ten days after receipt of the notification, and a conspicuous statement that unless the claim is paid within that time the goods will be advertised for sale and sold by auction at a specified time and place.

(d) The sale must conform to the terms of the notification.

(e) The sale must be held at the nearest suitable place to that where the goods are held or stored.

(f) After the expiration of the time given in the notification, an advertisement of the sale must be published once a week for two weeks consecutively in a newspaper of general circulation where the sale is to be held. Advertisement must include a description of the goods, the name of the person on whose account they are being held, and the time and place of the sale. The sale must take place at least fifteen days after the first publication. If there is no newspaper of general circulation where the sale is to be held, the advertisement must be posted at least ten days before the sale in not less than six conspicuous places in the neighborhood of the proposed sale.

(3) Before any sale pursuant to this section any person claiming a right in the goods may pay the amount necessary to satisfy the lien and the reasonable expenses incurred under this section. In that event, the goods must not be sold but must be retained by the warehouseman subject to the terms of the receipt and this Article.

(4) The warehouseman may buy at any public sale pursuant to this section.

(5) A purchaser in good faith of goods sold to enforce a warehouseman's lien takes the goods free of any rights of persons against whom the lien was valid, despite noncompliance by the warehouseman with the requirements of this section.

(6) The warehouseman may satisfy his lien from the proceeds of any sale pursuant to this section but must hold the balance, if any, for delivery on demand to any person to whom he would have been bound to deliver the goods.

(7) The rights provided by this section shall be in addition to all other rights allowed by law to a creditor against his debtor.

(8) Where a lien is on goods stored by a merchant in the course of his business, the lien may be enforced in accordance with either subsection (1) or (2).

(9) The warehouseman is liable for damages caused by failure to comply with the requirements for sale under this section and in case of willful violation is liable for conversion.

Part 3
Bills of Lading: Special Provisions

(Omitted)

Part 4
Warehouse Receipts and Bills of Lading: General Obligations

Sec. 7-401. Irregularities in Issue of Receipt or Bill or Conduct of Issuer.

The obligations imposed by this Article on an issuer apply to a document of title regardless of the fact that

(a) the document may not comply with the requirement of this Article or of any other law or regulation regarding its issue, form or content; or

(b) the issuer may have violated laws regulating the conduct of his business; or

(c) the goods covered by the document were owned by the bailee at the time the document was issued; or

(d) the person issuing the document does not come within the definition of warehouseman if it purports to be a warehouse receipt.

Sec. 7-402. Duplicate-Receipt or Bill; Overissue.

Neither a duplicate nor any other document of title purporting to cover goods already represented by an outstanding document of the same issuer confers any right in the goods, except as provided in the case of bills in a set, overissue of document for fungible goods and substitutes for lost, stolen or destroyed documents. But the issuer is liable for damages caused by his overissue or failure to identify a duplicate document as such by conspicuous notation on its face.

Sec. 7-403. Obligation of Warehouseman or Carrier to Deliver; Excuse.

(1) The bailee must deliver the goods to a person entitled under the document who complies with subsections (2) and (3), unless and to the extent that the bailee establishes any of the following:

(a) Delivery of the goods to a person whose receipt was rightful as against the claimant;

(b) damage to or delay, loss or destruction of the goods for which the bailee is not liable, (but the burden of establishing negligence in such cases is on the person entitled under the document);

Note: The brackets in (1)(b) indicate that State enactments may differ on this point without serious damage to the principle of uniformity.

(c) previous sale or other disposition of the goods in lawful enforcement of a lien or on warehouseman's lawful termination of storage;

(d) the exercise by a seller of his right to stop delivery pursuant to the provisions of the Article on Sales (Section 2-705);

(e) a diversion, reconsignment or other disposition pursuant to the provisions of this Article (Section 7-303) or tariff regulating such right;

(f) release, satisfaction or any other fact according a personal defense against the claimant;

(g) any other lawful excuse.

(2) A person claiming goods covered by a document of title must satisfy the bailee's lien where the bailee so requests or where the bailee is prohibited by law from delivering the goods until the charges are paid.

(3) Unless the person claiming is one against whom the document confers no right under Sec. 7-503(1), he must surrender for cancellation or notation of partial deliveries any outstanding negotiable document covering the goods, and the bailee must cancel the document or conspicuously note the partial delivery thereon or be liable to any person to whom the document is duly negotiated.

(4) "Person entitled under the document" means holder in the case of a negotiable document, or the person to whom delivery is to be made by the terms or pursuant to written instructions under a non-negotiable document.

Sec. 7-404. No Liability for Good Faith Delivery Pursuant to Receipt or Bill.

A bailee who in good faith including observance of reasonable commercial standards has received goods and delivered or otherwise disposed of them according to the terms of the document of title or pursuant to this Article is not liable therefore. This rule applies even though the person from whom he received the goods had no authority to procure the document or to dispose of the goods and even though the person to whom he delivered the goods had no authority to receive them.

Part 5

Warehouse Receipts and Bills of Lading; Negotiation and Transfer

Sec. 7-501. Form of Negotiation and Requirements of "Due Negotiation."

(1) A negotiable document of title running to the order of a named person is negotiated by his endorsement and delivery. After his endorsement in blank or bearer any person can negotiate it by delivery alone.

(2) (a) A negotiable document of title is also negotiated by delivery alone when by its original terms it runs to bearer.

(b) When a document running to the order of a named person is delivered to him the effect is the same as if the document had been negotiated.

(3) Negotiation of a negotiable document of title after it has been endorsed to a specified person requires endorsement by the special endorsee as well as delivery.

(4) A negotiable document of title is "duly negotiated" when it is negotiated in the manner stated in this section to a holder who purchases it in good faith without notice of any defense against or claim to it on the part of any person and for value, unless it is established that the negotiation is not in the regular course of business or financing or involves receiving the document in settlement or payment of a money obligation.

(5) Endorsement of a non-negotiable document neither makes it negotiable nor adds to the transferee's rights.

(6) The naming in a negotiable bill of a person to be notified of the arrival of the goods does not limit the negotiability of the bill nor constitute notice to a purchaser thereof of any interest of such person in the goods.

Sec. 7-502. Rights Acquired by Due Negotiation.

(1) Subject to the following section and to the provisions of Section 7-205 on fungible goods, a holder to whom a negotiable document of title has been duly negotiated acquires thereby:

(a) title to the document;

(b) title to the goods;

(c) all rights accruing under the law of agency or estoppel, including rights to goods delivered to the bailee after the document was issued; and

(d) the direct obligation of the issuer to hold or deliver the goods according to the terms of the document free of any defense or claim by him except those arising under the terms of the document or under this Article. In the case of a delivery order the bailee's obligation accrues only upon acceptance and the obligation acquired by the holder is that the issuer and any endorser will procure the acceptance of the bailee.

(2) Subject to the following section, title and rights so acquired are not defeated by any stoppage of the goods represented by the document or by surrender of such goods by the bailee, and are not impaired even though the negotiations or any prior negotiation constituted a breach of duty or even though any person has been deprived of possession of the document by misrepresentation, fraud, accident, mistake, duress, loss, theft or conversion, or even though a previous sale or other transfer of the goods or documents has been made to a third person.

Sec. 7-503. Document of Title to Goods Defeated in Certain Cases.

(1) A document of title confers no right in goods against a person who before issuance of the document had a legal interest or a perfected security interest in them and who neither

(a) delivered or entrusted them or any document of title covering them to the bailor or his nominee with actual or apparent authority to ship, store or sell or with power to obtain delivery under this Article (Section 7-403) or with power of disposition under this Act (Sections 2-403 and 9-307) or other statute or rule of law; nor

(b) acquiesced in the procurement by the bailor or his nominee of any document of title.

(2) Title to goods based upon an unaccepted delivery order is subject to the rights of anyone to whom a negotiable warehouse receipt or bill of lading covering the goods has been duly negotiated. Such a title may be defeated under the next section to the same extent as the rights of the issuer or a transferee from the issuer.

(3) Title to goods based upon a bill of lading issued to a freight forwarder is subject to the rights of anyone to whom a bill issued by the freight forwarder is duly negotiated; but delivery by the carrier in accordance with Part 4 of this Article pursuant to its own bill of lading discharges the carrier's obligation to deliver.

Sec. 7-504. Rights Acquired in the Absence of Due Negotiations; Effect of Diversion; Seller's Stoppage of Delivery.

(1) A transferee of a document, whether negotiable or non-negotiable, to whom the document has been delivered but not duly negotiated, acquired the title and rights which his transferor had or had actual authority to convey.

(2) In the case of a non-negotiable document, until but not after the bailee receives notification of the transfer, the rights of the transferee may be defeated

(a) by those creditors of the transferor who could treat the sale as void under Section 2-402; or

(b) by a buyer from the transferor in ordinary course of business if the bailee has delivered the goods to the buyer or received notification of his rights; or

(c) as against the bailee by good faith dealings of the bailee with the transferor.

(3) A diversion or other change of shipping instructions by the consignor in a non-negotiable bill of lading which causes the bailee not to deliver to the consignee defeats the consignee's title to the goods if they have been delivered to a buyer in ordinary course of business and in any event defeats the consignee's rights against the bailee.

(4) Delivery pursuant to a non-negotiable document may be stopped by a seller under Section 2-705, and subject to the requirement of due notification there provided. A bailee honoring the seller's instructions in entitled to be indemnified by the seller against any resulting loss or expense.

Sec. 7-505. Endorser Not a Guarantor for Other Parties.

The endorsement of a document of title issued by a bailee does not make the endorser liable for any default by the bailee or by previous endorsers.

Sec. 7-506. Delivery Without Endorsement: Right to Compel Endorsement.

The transferee of a negotiable document of title has a specifically enforceable right to have his transferor supply any necessary endorsement but the transfer becomes a negotiation only as of the time the endorsement is supplied.

Sec. 7-507. Warranties on Negotiation or Transfer of Receipt or Bill.

Where a person negotiates or transfers a document of title for value otherwise than as a mere intermediary under the next following section, then unless otherwise agreed, he warrants to his immediate purchaser only in addition to any warranty made in selling the goods

(a) That the document is genuine; and

(b) that he has no knowledge of any fact which would impair its validity or worth; and

(c) that his negotiation or transfer is rightful and fully effective with respect to the title to the document and the goods it represents.

Sec. 508. Warranties of Collecting Bank as to Documents.

A collecting bank or other intermediary known to be entrusted with documents on behalf of another or with collection of a draft or other claim against delivery of documents warrants by such delivery of the documents only its own good faith and authority. This rule applies even though the intermediary has purchased or made advances against the claim or draft to be collected.

Sec. 7-509. Receipt or Bill: When Adequate Compliance with Commercial Contract.

The question whether a document is adequate to fulfil the obligations of a contract for sale or the conditions of a credit is governed by the Articles on Sales (Article 2) and on Letters of Credit (Article 5).

Part 6

Warehouse Receipts and Bills of Lading: Miscellaneous Provisions

Sec. 7-601. Lost and Missing Documents.

(1) If a document has been lost, stolen or destroyed, a court may order delivery of the goods or issuance of a substitute document and the bailee may without liability to any person comply with such order. If the document was negotiable, the claimant must post security approved by the court to indemnify any person who may suffer loss as a result of non-surrender of the document. If the document was not negotiable, such security may be required at the discretion of the court. The court may also in its discretion order payment of the bailee's reasonable costs and counsel fees.

(2) A bailee who without court order delivers goods to a person claiming under a missing negotiable document is liable to any person injured thereby, and if the delivery is not in good faith, becomes liable for conversion. Delivery in good faith is not conversion if made in accordance with a filed classification or tariff or, where no classification or tariff is filed, if the claimant posts security with the bailee in an amount at least double the value of the goods at the time of posting to indemnify any person injured by the delivery who files a notice of claim within one year after the delivery.

Sec. 7-602. Attachment of Goods Covered by a Negotiable Document.

Except where the document was originally issued upon delivery of goods by a person who had no power to dispose of them, no lien attaches by virtue of any judicial process to goods in the possession of a bailee for which a negotiable document of title is outstanding unless the document be first surrendered to the bailee or its negotiation enjoined, and the bailee shall not be compelled to deliver the goods pursuant to process until the document is surrendered to him or impounded by the court. One who purchases the document for value without notice of the process of injunction takes free of the lien imposed by judicial process.

Sec. 7-603. Conflicting Claims; Interpleader.

If more than one person claims title or possession of the goods, the bailee is excused from delivery until he has had a reasonable time to ascertain the validity of the adverse claims or to bring an action to compel all claimants to interplead and may compel such interpleader, either in defending an action for non-delivery of the goods, or by original action, whichever is appropriate.

CHAPTER 12

PAYMENT OF TRANSPORTATION CHARGES

Section 10743 of P.L. 95-473 reads as follows:

> "(a) Except as provided in subsection (b) of this section, a common carrier (except a pipeline or sleeping car carrier) providing transportation or service subject to the jurisdiction of the Interstate Commerce Commission under this subtitle shall give up possession at destination of property transported by it only when payment for the transportation or service is made.
>
> (b) (1) Under regulations of the Commission governing the payment for transportation and service and preventing discrimination, those carriers may give up possession at destination of property transported by them before payment for the transportation or service. The regulations of the Commission may provide for weekly or monthly payment for transportation provided by motor common carriers and for periodic payment for transportation provided by water common carriers.
>
> (2) Such a carrier (including a motor common carrier being used by a freight forwarder) may extend credit for transporting property for the United States Government, a State, a territory or possession of the United States, or a political subdivision of any of them.

Prior to the Transportation Act of 1920, the Act was silent as to the time within which freight charges should be paid. Failure of shippers to pay charges promptly had the effect of withholding from the carriers working capital of many millions of dollars daily. Shippers were thus improperly making use of capital that should have been turned over to the carriers in payment of transportation charges. Many shippers obtained long-time credit while others were required to make prompt payment. This created unjust discrimination, since prompt payment was required in some cases and long-time credit extended in others.

While this practice was given consideration from time to time with a view of establishing a uniform period under which transportation charges should be paid, nothing of a decisive nature was agreed upon until the railroads were placed under Federal control during World War I. The Director General of Railroads, under General Order No. 25 ordered that, effective August 1, 1918, the collection of transportation charges by all carriers under Federal control was to be on a cash basis. Credit accommodations that were in connflict with this order were ordered to be cancelled. An exception was provided, however, whereby credit not exceeding 48 hours could be granted provided the shipper filed a surety bond that was satisfactory to the carriers.

This matter was given consideration in the drafting of the Transportation Act of 1920 and Congress added paragraph 2 to Section 3 of the Interstate Commerce Act at that time.

Formal credit regulations were originally prescribed by the Commission in *Ex Parte 73,* (57 I.C.C. 591) and several modifications have been made since that time. The current credit regulations are published in *Chapter X, Title 49 of the Code of Federal Regulations.* They are clear and concise and leave no doubt as to their application and yet each year numerous shippers and carriers are fined for violations of the credit regulations. Since there is no need for explanation or interpretation the credit rules are reproduced as they appear in the *Code of Federal Regulations.*

PARTS 1320-1329 CREDIT REGULATIONS

Part 1320—Extention of Credit to Shippers By Rail Carriers

"1320.1 Carrier may extend credit to shipper.

"The carrier, upon taking precautions deemed by it to be sufficient to assure payment of the tariff charges within the credit periods specified in this part, may relinquish possession of freight in advance of the payment of the tariff charges thereon and may extend credit in the amount of such charges to those who undertake to pay such charges, such persons hereon being called shippers, for a period of 4 days (or 5 days where retention or possession of freight by the carrier until tariff rates and charges thereon have been paid will retard prompt delivery or will retard prompt release of equipment or station facilities) as set forth in this part. In regard to traffic of nonprofit shippers' associations and shippers' agents, within the meaning of section 402(c) of part IV of the Interstate Commerce Act, the carriers shall require such organizations to furnish the names of the beneficial owners of the property in the bills of lading or at least have the bills of lading incorporate by reference a document containing the names of the beneficial owners.

"1320.2, 1320.3, and 1320.4— [Reserved]

"1320.5 Additional charges.

"Where a carrier has relinquished possession of freight and collected the amount of tariff charges represented in a freight bill presented by it as the total amount of such charges, and another freight bill for additional charges is thereafter presented to the shipper, the carrier may extend credit in the amount of such additional charges for a period of 30 days, to be computed as set forth in this part, from the date of the presentation of the subsequently presented freight bill.

"Cross Reference: For presentation of freight bills and periods of credit following, see 1320.9-1320.11.

"1320.6 Icing charges.

"Where icing charges are not published in the tariffs at fixed amounts determinable at the time the shipment moves from point of origin, and where freight charges are prepaid and icing charges are to be paid by the consignor, the carrier, upon taking precautions deemed by it to be sufficient to assure prompt payment of the tariff charges within the credit area specified in this part, may relinquish possession of the freight in advance of the payment of the icing charges and may delay presentation of bills for such icing charges for a period not exceeding 15 days after the end of the calendar month during which the charges accrued and may extend credit in the amount of such charges for 15 days from the presentation of the bill for such charges.

"1320.7 Demurrage charges.

"Where the amount of demurrage charges is determinable under average agreements made in accordance with tariff provisions, the carrier, upon taking precautions deemed by it to be sufficient to assure prompt payment of the tariff charges within the credit period, may delay the presentation of bills for such demurrage charges for a period not to exceed 15 days from the expiration of the authorized demurrage period and may extend credit in the amount of the demurrage charges accruing during the demurrage period for 15 days from the presentation of the bill for such charges.

"1320.8 Periods of credit following delivery.

"Where the freight bill is presented to the shipper prior to, or at the time of, delivery of the freight, the 4- and 5-day periods of credit shall run from the first 12 o'clock midnight following the delivery of the freight.

"1320.9 Periods of credit following presentation of freight bill.

"Where the freight bill is presented to the shipper subsequent to the time the freight is delivered, the 4- and 5-day periods of credit shall run from the first 12 o'clock midnight following the presentation of the freight bill.

"1320.10 Presentation of freight bills.

"Every carrier shall present freight bills for all transportation charges except those specifically excepted in this part to shippers prior to the first 12 o'clock midnight following forwarding of prepaid shipments or delivery of collect shipments except that when information sufficient to enable the carrier to compute the tariff charges is not then available to the carrier at the billing point, the freight bills shall be presented not later than the first 12 o'clock midnight following the day upon which sufficient information becomes available at the billing point of the carrier. A carrier shall not extend further credit to any shipper which fails to furnish sufficient information to allow the carrier to render a freight bill within a reasonable time after the shipment is tendered to the origin carrier. As used in this section the term "shipper" includes, but is not limited, to, freight forwarders as well as shippers' associations and shippers' agents within the meaning of section 402(c) of part IV of the Interstate Commerce Act.

"1320.11 Time of presentation when mailed.

"Shippers may elect to have their freight bills presented by means of the United States mails, and when the mail service is so used the time of mailing by the carrier shall be deemed to be the time of presentation of the bills. In case of dispute as to the time of mailing the postmark shall be accepted as showing such time.

"1320.12 Saturdays, Sundays, and legal holidays may be excluded from period.

"In the computation of the various periods of credit Saturdays, Sundays, and legal holidays may be excluded, and where the time for presentation to shippers of freight bills for transportation and related charges falls on Saturday, Sunday or a legal holiday such bills may be presented prior to 12 o'clock midnight of the next succeeding regular work day.

"1320.13 Time of collection when payment is mailed.

"The mailing by the shipper of valid checks, drafts, or money orders, which are satisfactory to the carrier, in payment of freight charges within the credit periods allowed such shipper may be deemed to be the collection of the tariff charges within the credit period for the purposes of the rules in this part. In case of dispute as to the time of mailing the postmark shall be accepted as showing such time.

"1320.14 Computation of 96-hour period for payment of demurrage charges on coal delivered to tidewater exchanges.

"The period of 96 hours fixed for the payment of transportation rates and charges, insofar as applicable to the payment of demurrage accruing at New York, N.Y., Philadelphia, Pa., Baltimore, Md., and Sewalls Point (Norfolk), Va., on coal pooled for transshipment at those points under control of coal exchanges, may be computed from the first 4 p.m. following the day when the demurrage bills are presented by such exchanges to individual shippers, members thereof.

"1320.15 Computation of 96-hour period for payment of export traffic rates.

"The period of 96 hours fixed for the payment of transportation rates and charges, insofar as applicable to export traffic which is loaded into vessels direct from railroad cars or piers or from such cars or piers by means of lighters, may be computed from the first 4 p.m. following the time when the vessel is completely loaded, freight bills to be delivered to vessel owner or his representative not later than the day on which the loading of the vessel is completed.

"1320.16 Computation of 96-hour period for payment of freight rates at interior California points not served by railroads.

"The period of 96 hours fixed for the payment of transportation rates and charges, insofar as applicable to freight from and to Bartle, Calif., when destined to or from interior points described in the report, in Siskiyou, Shasta, and Modoc Counties, Calif., not served by railroad, may be computed from the first 4 p.m. following 26 days after the mailing by petitioner of the freight bills for such traffic.

"1320.17 Interline settlement of revenues.

"Nothing in this part shall be interpreted as affecting the interline settlement of revenues from traffic which is transported over through routes composed of lines of common carriers subject to parts I, II, or III of the Interstate Commerce Act.

Part 1321—Extension of Credit to Shippers By Express Companies

"1321.1 Credit period allowed; Saturdays, Sundays, and legal holidays may be excluded.

"Upon taking precautions deemed by them to be sufficient to assure payment of the tariff charges within the credit period herein specified, express companies subject to Part I of the Interstate Commerce Act may relinquish possession of express in advance of the payment of the tariff charges thereon and may extend credit in the amount of such charges to those who undertake to pay them, such persons herein being called shippers, for a period of 7 days, excluding Saturdays, Sundays, and legal holidays, computed as hereinafter set forth. The credit period shall run from the first 12 o'clock midnight following the presentation of the express bill.

"1321.2 Additional charges.

"Where an express company has relinquished possession of express and collected the amount of the tariff charges represented in an express bill presented by it as the total amount of such charges, and another express bill for additional express charges is thereafter presented to the shipper, the express company may extend credit in the amount of such additional charges for a period of 30 calendar days to be computed from the first 12 o'clock midnight following the presentation of the subsequently presented express bill.

"1321.3 Presentation of express bills.

"Express bills for all transportation charges may cover all transactions, collect shipments delivered and prepaid shipments picked up, handled during a calendar week, designated as the billing week. Express bills for all transportation charges shall be presented to the shippers within 4 working days following the close of the billing week. When mail service is used, the time of mailing by the carrier shall be deemed to be the time of presentation of the bills. In case of dispute as to the time of mailing, the postmark shall be accepted as showing such time.

"1321.4 Time of collection when payment is mailed.

"The mailing by the shipper of valid checks, drafts or money orders, which are satisfactory to the express company, in payment of express charges within the credit period allowed such shipper may be deemed to be the collection of the tariff rates and charges within the credit period for the purpose of these rules. In case of dispute as to the time of mailing, the postmark shall be accepted as showing such time.

Part 1322—Extension of Credit to Shippers By Motor Carriers

"1322.1 Carrier may extend credit to shipper.

"(a) Extension of credit. Upon taking precautions deemed by them to be sufficient to assure payment of the tariff charges within the credit period herein specified, common carriers by motor vehicle may relinquish possession of freight in advance of the payment of the tariff charges thereon and may extend credit in the amount of such charges to those who undertake to pay them, such persons herein being called shippers, for a period of 7 days excluding Saturdays, Sundays, and legal holidays. When the freight bill covering a shipment is presented to the shipper on or before the date of delivery, the credit period shall run from the first 12 o'clock midnight following delivery of the freight. When the freight bill is not presented to the shipper on or before the date of delivery, the credit period shall run from the first 12 o'clock midnight following the presentation of the freight bill. In regard to traffic of non-profit shippers' associations and shippers'

agents, within the meaning of section 402(c) of part IV of the Interstate Commerce Act, the carriers shall require such organizations to furnish the names of the beneficial owners of the property in the bills of lading or at least have the bills of lading incorporate by reference a document containing the names of the beneficial owners.

"*(b)*— [Reserved]

"*(c) Exceptions—Carriers of household goods.* The provisions of the paragraph (a) of this section shall not apply in any instance in which the carrier shall be required by § 1056.8 (b) of this chapter, "Transportation of household goods in interstate or foreign commerce." to relinquish possession of a shipment of household goods in advance of the payment of the total amount of the tariff charges thereon.

"*(d) Exceptions—Carriers of household goods.* Except as provided in paragraph (b) of this section, motor common carriers of household goods must also provide in their tariffs that (1) the aforestated credit period of seven days excluding Saturdays, Sundays, and legal holidays shall automatically be extended to a total of 30 calendar days for any shipper who has not paid the carrier's freight bill within the aforesaid seven-day period, (2) such shipper will be assessed a service charge by the carrier equal to one per cent of the amount of said freight bill, subject to a $10 minimum charge, for such extension of the credit period, and (3) no such carrier shall grant credit to any shipper which fails to pay a duly presented freight bill within the 30-day period, unless and until such shipper affirmatively satisfies the carrier that all future freight bills duly presented will be paid strictly in accordance with the rules and regulations prescribed by the Commission for the settlement of carrier rates and charges. *Provided,* That no service charge authorized herein shall be assessed in connection with rates and charges on freight transported for the United States, for any department, bureau, or agency thereof, or for any State or territory, or political subdivision thereof, or for the District of Columbia.

"*1322.2 Credit for additional charges after freight relinquished.*

"Where a common carrier by motor vehicle has relinquished possession of freight and collected the amount of tariff charges represented in a freight bill presented by it as the total amount of such charges, and another freight bill for additional charges is thereafter presented to the shipper, the carrier may extend credit in the amount of such additional charges for a period of 30 calendar days, to be computed from the first 12 o'clock midnight following the presentation of the subsequently presented freight bill.

"*1322.3 Period of credit following delivery of freight.*

"Freight bills for all transportation charges shall be presented to the shippers within seven calendar days from the first 12 o'clock midnight following delivery of the freight except that motor common carriers of household goods and motor common carriers of oilfield equipment shall present their freight bills for all transportation charges to the shipper within 15 calendar days, excluding Saturdays, Sundays, and holidays, from the first 12 o'clock midnight following delivery of the freight.

"1322.4 Freight bills may be presented by mail.

"Shippers may elect to have their freight bills presented by means of the United States mails, and when the mail service is so used the time of mailing by the carrier shall be deemed to be the time of presentation of the bills. In case of dispute as to the time of mailing, the postmark shall be accepted as showing such time.

"1322.5 Payment by checks, drafts or money orders.

"The mailing by the shipper of valid checks, drafts, or money orders, which are satisfactory to the carrier, in payment of freight charges within the credit period allowed such shipper, may be deemed to be the collection of the tariff charges within the credit period for the purposes of the rules in this part. In case of dispute as to the time of mailing, the postmark shall be accepted as showing such time.

Part 1323—Settlement of Rates and Charges of Common Carriers By Water

"1323.1 Relinquishment of freight in advance of payment of charges.

"All common carriers of propery by water subject to Part III of the Interstate Commerce Act, after having taken precautions deemed by them to be sufficient to assure payment of their freight charges within the credit periods hereinafter specified, such as examination of the credit rating of the person or persons undertaking to pay the freight charges or the obtaining of satisfactory surety bonds, are hereby authorized to relinquish possession of freight at destination or in advance of the payment of the tariff charges lawfully due thereon and to extend credit to those who undertake to pay such charges, as hereinafter authorized.

"1323.2 Extension of credit for 48-hour period.

"Except as otherwise provided in 1323.3, credit may be extended for a period not exceeding 48 hours, computed as provided in 1323.4.

"1323.3 Extension of credit for 96-hour period.

"When retention of possession of freight by the carrier until the tariff rates and charges thereon have been paid will retard prompt delivery or will retard prompt release of equipment or terminal facilities of the carrier, credit may be extended for a period not exceeding 96 hours, computed as provided in 1323.4.

"1323.4 Computation of credit period.

"(a) When the freight bill is presented to the person or persons undertaking to pay the charges prior to, or at the time of, delivery of the freight, the 48-hour and 96-hour periods of credit shall run from the first 12 o'clock midnight following delivery of the freight.

"(b) When the freight bill is presented to the person or persons undertaking to pay the charges subsequent to the time the freight is delivered, the 48-hour and 96-hour periods of credit shall run from the first 12 o'clock midnight following presentation of the freight bill.

"(c) In the computation of the various periods of credit Saturdays, Sundays and legal holidays may be excluded, and where the time for presentation of freight bills for transportation and related charges falls on Saturdays, Sunday or a legal holiday such bills may be presented prior to 12 o'clock midnight of the next succeeding regular work day.

"1323.5 Presentation of freight bills; mailing.

"(a) Except as otherwise provided in paragraph (b) of this section and in §§ 1323.4, 1323.6, and 1323.7, carriers shall present freight bills for all transportation charges to the person or persons undertaking to pay those charges as promptly as practicable but in every case prior to the second 12 o'clock midnight following delivery of the freight.

"(b) When information sufficient to enable the carrier to compute the tariff charges is not available to the carrier at the point where it computes the charges, presentation of the freight bill may be delayed until such information is available. In such cases it shall be the duty of the shipper (or consignee, as the case may be) to present, and of the carrier to obtain, the information as promptly as practicable. If, in any case, the necessary information has not become available to the carrier at the point where it computes the charges within 15 days after delivery of the freight, carrier shall present the freight bill and collect charges based upon the best information in its possession and arrange for correction later when detailed information is furnished.

"(c) The person or persons undertaking to pay freight charges may elect to have their freight bills presented by means of the United States mails. When mail service is so used, the time of mailing by the carrier shall be deemed to be the presentation of the bills. In case of dispute as to the time of mailing, the postmark shall be accepted as showing such time.

"1323.6 Extension of credit for additional charges.

"Where carrier has relinquished possession of freight and collected the amount of tariff charges represented in a freight bill presented by it as the total amount of such charges and another freight bill for additional charges is thereafter presented to the shipper the carrier may extend credit in the amount of such additional charges for a period of 30 days, to be computed as herein set forth, from the date of the presentation of the subsequently presented freight bill.

"1323.7 Extension of credit for demurrage charges.

"Where the amount of demurrage charges is determinable under average agreements made in accordance with tariff provisions the carrier may delay the presentation of bills for such demurrage charges for a period not to exceed 15 days from the expiration of the authorized demurrage period and may extend credit in the amount of the demurrage period for 15 days from the presentation of the bill for such charges.

"1323.8 Collection of freight charges; mailing.

"Mailing by the person or persons paying the freight charges of valid checks, drafts or money orders, which are satisfactory to the carrier, in payment of freight charges within the credit periods allowed, may be deemed to be the collection of the tariff charges within the credit period. In case of dispute as to the time of mailing, the postmark shall be accepted as showing such time.

Part 1324—Settlement of Freight Charges By Freight Forwarders

"1324.1 Credit for freight charges on forwarder freight by freight forwarders subject to part IV of the Interstate Commerce Act and by common carriers by motor vehicles subject to part II of that act collecting charges for such forwarders.

"The rules and regulations prescribed in Part 1322 of this chapter, governing the extension of credit for freight charges by common carriers by motor vehicle in case of service subject to the rules and regulations prescribed in Part 1322 of this chapter, governing the extension of credit for freight charges by common carriers by motor vehicle in case of service subject to part II of the Interstate Commerce Act, shall apply on and after August 14, 1942, to freight forwarders subject to part IV of the Interstate Commerce Act and to common carriers by motor vehicle subject to part II of that act when collecting charges for such forwarders, in the case of service subject to part IV of that act.

CHAPTER 13

OVERCHARGES AND UNDERCHARGES

Charges to Be Paid In Money and In Money Only

Section 10761 of Public Law 95-473 provides in part that:

> "That carrier may not charge or receive a different compensation for that transportation or service than the rate specified in the tariff."

The above provision applies to all carriers subject to the jurisdiction of the Interstate Commerce Commission under chapter 105 of Title 49. Also in connection with the collection of freight charges Section 10762(a)(2) requires motor carriers, water carriers and freight forwarders to state their rates in money of the United States. It is not permissible for a carrier to accept services, merchandise, or any other payment in lieu of the amount of money specified in the published rate schedules.

In the case of *L. & N. Railroad Company v. Mottley,* (219 U.S. 467, 476), the United States Supreme Court, in passing upon this feature said:

> "In our opinion, after the passage of the commerce act, the railroad company could not lawfully accept from Mottley and wife any compensation 'different' *in kind* from that mentioned in its published schedule of rates. And it cannot be doubted that the rates or charges specified in such schedule were payable only in money. They could not be paid in any other way, without producing the utmost confusion and defeating the policy established by the acts regulating commerce. The evident purpose of Congress was to establish uniform rates for transportation, to give all, the same opportunity to know what the rates were as well as to have the equal benefit of them. To that end the carrier was required to print, post and file its schedules and to keep them open to public inspection. No change could be made in the rates embraced by the schedule except upon notice to the Commission and to the public. But an examination of the schedules would be of no avail and would not ordinarily be of any practical value if the published rates could be disregarded in special or particular cases by the acceptance of property of various kinds, and of such value as the parties immediately concerned chose to put upon it, in place of money for services performed by the carrier. *** The passenger has no right to buy tickets with services, advertising, releases or property, nor can the railroad buy services, advertising, releases or property with transportation. The statute manifestly means that the purchase of a transportation ticket by a

passenger and its sale by the company shall be consummated only by the former paying cash and by the latter receiving cash of the amount specified in the published tariffs."

Counterclaims Not Permitted

It is also well settled, as an analogous principle to the fact that a carrier can accept only money in settlement of freight charges, that a carrier may not offset damages for injuries to goods against any part or all of the freight due on a shipment. Were carriers permitted to do this, they would be accepting damaged merchandise in payment of transportation charges which is forbidden by the Act. A carrier must collect its full tariff rates and charges and may not make any compromise in respect to the freight charges legally applicable.

In this connection, the United States District Court, in the case of *I.C. R.R. Co. v. W.L. Hoopes & Son et. al.,* (223 Fed. Rep. 135, 136), in which it quoted *C. & N.W. Ry. Co. v. Stein Co.,* (233 Fed. Rep. 716), Judge Munger presiding stated:

> "A few years ago the nation was startled when it was found that favorite shippers were getting rich upon the rebates granted by carriers. One of the devices used in granting such rebates was for the shipper to make a fictitious claim for damages to freight, which was promptly allowed. To permit counterclaims in these actions for freight charges would simply open the door to a renewal of this method of rebate; at least, it would cast suspicion upon the transactions—especially in every case where a compromise was effected. Compromises are favored in law. If the defendant in this action has a valid claim for damages, both parties should be permitted to exercise their rights to compromise the action; but, if such compromise were effected in a transaction involving the collection of freight charges, the court would be compelled to supervise it with the utmost care, in order, as Judge Munger expresses it 'to prevent the granting and receiving of rebates by insidious agreement between the parties.' "

In *Akers Motor Lines v. Lade Cornell Comb Co.,* (203 F. Supp. 156), the court held:

> "Plaintiff motor carrier has not only the right but the duty to collect proper charges for transportation services performed. While consignee who assumed responsibility for payment of charges in issue, is entitled to recover for damages to its goods due to carrier's negligence or misconduct, existence of such damage claim, even though valid, is no defense to plaintiff's claim for transportation charges."

Overcharges and Undercharges Prohibited By the Act

As the law specifically provides that no carrier shall charge or receive a *different compensation* for transporting freight than that named in the tariffs lawfully applicable at the time the shipment moves, it is quite evident that any deviation from the tariff through the charging of more than is properly assessable would be an overcharge against the shipper, while the charging of less than is covered by tariff provision would be an undercharge.

An overcharge or an undercharge may come about in many different ways. It may be the result of a simple error in classification, use of wrong tariff, improper description or billing, such as failure to bill different articles or parts separately, error in extension, advance charges, errors in weights, the use of a car with minimum capacity exceeding that of the car ordered, or failure to properly apply the rule preventing the assessing of a higher total charge for less-than-carload than for carload shipments, mixed carload rules, dunnage rules, or other rules in the classification and tariffs of the carriers. Whatever the cause however, the result is brought about by a deviation from the lawful tariff provisions and should be dealt with on that basis, by supplying the necessary tariff authority to support the position taken.

As both overcharges and undercharges are prohibited, the former because of their unreasonableness and the latter because of the discriminatory feature and the fostering of rebates, the carrier must adjust any inconsistencies in tariff application which are brought to its attention.

It is, of course, appreciated that because of the complexity of tariffs and their application it is practically impossible to expect that the correct charges will be assessed in every instance. However, if and when an undercharge is detected by the carrier, or is called to its attention, it must exhaust every means at its command to effect collection of the proper tariff charges, even to the extent of filing suit and taking any other legal action which may be necessary in the premises. In the case of overcharges the carrier must refund to the shipper or consignee, according to who pays the freight charges, anything in excess of the legally applicable tariff charges. In the event the carrier refuses to do so, because of unwillingness or inability to agree to the interpretation of the claimant, then the aid of the Interstate Commerce Commission or the courts may be invoked to settle the controversy.

Parties Liable for Freight Charges

A claim for overcharges may be made by the shipper or consignee, according to who may have paid the freight charges on the shipment. It is, of course, collectible from the carrier which issued the expense bill and received the overcharge. In the case of a prepaid shipment this would be the initial line and in the case of a collect shipment, the delivering carrier.

When an undercharge occurs the carrier is at liberty to proceed against either the consignee or the consignor. If the shipment has been prepaid it will usually look to the consignor for the amount of the balance due and if the shipment has been sent collect will request the consignee to pay the full tariff charges. In any event the law places upon the carrier an absolute obligation to collect the lawfully published freight charges on an interstate shipment and it may not waive any undercharge to which it is legally entitled unless all means of proceeding against either the consignor or the consignee, or both, have been employed. This is true even to the extent that if a carrier, through oversight or for any other reason, delivers a collect shipment to a consignee and it subsequently develops that it is unable to collect all or part of its charges from such consignee, then the consignor must be looked to for satisfaction of the lawful freight charges due on the shipment.

While the Supreme Court of the United States has held in the case of *P.C.C. & St. L. Ry. Co. v. Alvin J. Fink,* (250 U.S. 577), that the consignee, by acceptance of the goods, becomes liable for the full amount of the freight charges, if the carrier finds that, because of bankruptcy or other financial difficulties of the consignee, it is unable to collect such charges, then they must be paid by the consignor. Inasmuch as there rests upon the carrier the obligation to collect the legally applicable freight charges, there is a reciprocal obligation of equal extent upon either the shipper or the consignee to pay such charges and the carrier's right to collect from the consignee, as above referred to, does not imply a release of the consignor.

In connection with this feature, the Circuit Court of Appeals of Georgia, Division No. 1, in the case of *Southern Railway Company v. Southern Cotton Oil Company,* (91 S.E. 876), held as follows:

> "A railroad company which, through mistake or negligence, has failed to collect from a consignee the charges due for transportation, is not estopped from recovering them from the consignor, merely because of failure to sue therefore until after the consignee (who by agreement with the consignor is liable for the freight) has become insolvent."

This finding of the court has for its basis the fact that it is the consignor of the shipment who contracts with the carrier for its carriage, and is, therefore, liable for payment of the transportation charges. While the consignee, by acceptance of the goods, may become liable for the freight charges, still that does not invalidate the contract of carriage which the consignor entered into.

Nonrecourse Clause

The previous statement may, however, be qualified by the assertion that in Section 7 of the uniform bill of lading the shipper is given the option to indicate on the face of the bill of lading, in the space arranged for that purpose, that the shipment is to be delivered to the consignee without recourse on the consignor. This has the effect of notifying the carrier that, as part of the contract of carriage, the consignor expects it to collect all charges from the consignee. Under the ruling of the Supreme Court of the United States in the case of *L. & N. R.R. Co. v. Central Iron & Steel Company,* (265 U.S. 59, 66), if the carrier should, in the face of this stipulation, make delivery to the consignee without requiring full payment of the freight charges, the shipper could not be held liable.

Another ruling of the Supreme Court, of interest in this connection, was made in *Illinois steel Co. v. Baltimore & Ohio R. Co.,* (320 U.S. 508). In this case, the Supreme Court reversed a decision of a State court in Illinois, 316 Ill. App. 516, involving a construction of the uniform railroad bill of lading approved by the Interstate Commerce Commission.

The petitioner had made shipments over respondent's railroad of certain freight intended to be exported, and had prepaid the freight charges based on the export rail rates. At the rail destination the shipments were so handled by the consignee as to make higher domestic rates applicable, and respondent undertook to collect undercharges from the petitioner. The bill of lading contained a "non-recourse clause" signed by the petitioner providing that the carrier should not make delivery of the shipment without payment of freight and all other lawful charges. The question presented was whether a stipulation in the bill of lading for the prepayment of freight by the consignor restricted the operation of the nonrecourse clause, so that, despite its presence in the bill of lading, recourse may be had to the consignor for charges in addition to those which were prepaid at shipment, the additional charges arising only by reason of events which occurred on or after the delivery of the shipments to the consignee. The provision in question exempts the consignor from liability for any undercharges when the carrier makes delivery without requiring payment by the consignee of any such undercharges.

The Court concluded "that the reasonable construction of the prepayment clause is that, with respect to these charges, it did not either by its design or by the intention of the parties, curtail the operation of the nonrecourse clause so as to deprive the petitioner, the consignor, of the immunity from liability for which it was entitled to stipulate by the nonrecourse clause."

Liability of Beneficial Owner

Another outstanding exception to the general rule of liability for freight charges, on the part of the consignee, as laid down in the *Fink Case, supra,* is witnessed in the provisions of Section 3, paragraph 2, of the Interstate Commerce Act, which became law in the so-called *Newton Bill* of March 4, 1927, and is now contained in Section 10744 of P.L. 95-473, as follows:

"(a)(1) Liability for payment of rates for transportation for a shipment of property by a shipper or consignor to a consignee other than the shipper or consignor, is determined under this subsection when the transportation is provided by a rail, motor, or water common carrier under this subtitle. When the shipper or consignor instructs the carrier transporting the property to deliver it to a consignee that is an agent only, not having beneficial title to the property, the consignee is liable for rates billed at the time of delivery for which the consingee is otherwise liable, but not for additional rates that may be found to be due after delivery if the consignee gives written notice to the delivering carrier before delivery of the property—

"(A) of the agency and absence of beneficial title; and

"(B) of the name and address of the beneficial owner of the property if it is reconsigned or diverted to a place other than the place specified in the original bill of lading.

"(2) When the consignee is liable only for rates billed at the time of delivery under paragrph (1) of this subsection, the shipper or consignor, or, if the property is reconsigned or diverted, the beneficial owner, is liable for those additional rates regardless of the bill of lading or contract under which the property was transported. The beneficial owner is liable for all rates when the property is reconsigned or diverted by an agent but is refused or abandoned at its ultimate destination if the agent gave the carrier in the reconsignment or diversion order a notice of agency and the name and address of the beneficial owner. A consignee giving the carrier, and a reconsignor or diverter, giving a rail carrier erroneous information about the identity of the beneficial owner of the property is liable for the additional rates.

"(b) Liability for payment of rates for transportation for a shipment of property by a shipper or consignor, named in the bill of lading as consignee, is determined under this subsection when the transportation is provided by a rail or express carrier under this subtitle. When the shipper or consignor gives written notice, before delivery of the property, to the line-haul carrier that is to make ultimate delivery—

"(1) to deliver the property to another party identified by the shipper or consignor as the beneficial owner of the property; and

"(2) that delivery is to be made to that party on payment of all applicable transportation rates;

that party is liable for the rates billed at the time of delivery and for additional rates that may be found to be due after delivery if that party does not pay the rates required to be paid under clause (2) of this subsection on delivery. However, if the party gives written notice to the delivering carrier before delivery, that the party is not the beneficial owner of the property, and gives the carrier the name and address of the beneficial owner, then the party is not liable for those additional rates. A shipper, consignor, or party to whom delivery is made that gives the delivering carrier erroneous information about the identity of the beneficial owner, is liable for the additional rates regardless of the bill of lading or contract under which the property was transported. This subsection does not apply to a prepaid shipment of property.

"(c)(1) A rail carrier may bring an action to enforce liability under subsection (a) of this seciton. That carrier must bring the action during the period provided in section 11706(a) of this title or by the end of the 6th month after final judgment against it in an action against the consignee, or the beneficial owner named by the consignee or agent, under that section.

"(2) A water common carrier may bring an action to enforce liability under subsection (a) of this section. That carrier must bring the action by the end of the 2nd year after the claim accrues or by end of the 6th month after final judgment against it in an action against the consignee or beneficial owner named by the consignee by the end of that 2-year period.

"(3) A rail or express carrier may bring an action to enforce liability under subsection (b) of this section. That carrier must bring the action during the period provided in section 11706(a) of this title or by the end of the 6th month after final judgment against it in an action against the shipper, consignor, or other party under that section."

The purpose of this amendment is to protect commission merchants and agents, who handle shipments for the accounts of shippers or consignors mostly on a consignment or commission basis, so that they will not be called upon by the carriers to pay undercharges on shipments after they have made remittance of the proceeds to their principals and the account has been closed. For example, under the liability imposed upon the consignee in accordance with the holding of the Superior Court in the *Fink case, supra,* the consignee, in the absence of the above provisions could be held liable for the entire amount of the freight charges. This would hold true even though the carrier, through error, had originally presented a freight bill to the consignee for an amount less than the legal freight charges, which error it corrected several months later through the issuance of an undercharge bill, the consignee in the meantime having closed out that particular account and remitted the proceeds to the shipper on the basis of the freight charges as contained in the original bill. In situations of this kind it often happened that by the time the consignee paid the due bill of the carrier the transaction resulted in a loss instead of a profit. Since the shipper could not always be located, the result was that the amount of the undercharge had to be borne wholly by the consignee. Under the present provisions, however, the consignee is enabled to guard against this possibility, through proper advice to the carrier prior to the delivery of the property that he has no beneficial title in the goods. This puts the carrier on guard in seeing that the proper tariff charges are collected in the first instance.

In this connection, the Federal court in *New York Central R. Co. v. Transamerican Petroleum Corp.*, (108 Fed. (2d) 944, 998), held:

> "Written directions by the consignee to the carrier to deliver the freight to a third person from whom collection was to be made was an offer which the carrier, by its act of performance, accepted, and thus a contractual relation resulted."

As to the term "beneficial title" as used in this provision of the statute, there has not, as yet, been a judicial definition of just what this embraces. It seems apparent, however, that, this term, when read in conjunction with its companion term "beneficial owner," as used therein, is designed to distinguish the legal title to the goods which is vested in the owner of such goods, as contrasted with that interest which a commission merchant or an agent has in goods which are consigned to him for sale.

Consignee Not Liable for Charges on Refused Shipments

It should not be assumed from the foregoing that the consignee may be held liable for the freight charges in all instances other than where the provisions of Section 10744, are complied with. Even if the consignor were to sign Section 7 of the bill of lading, as previously referred to, the consignee could still escape liability for payment of the freight charges if the circumstances were such that he found it necessary to refuse the shipment.

This is accounted for by the fact that there is no contractual relation between the carrier and the consignee (or the order notify party in the case of an order notify shipment) by mere designation of such party as consignee or order notify party, which obligates him to receive the goods or to pay the freight charges, and that party is not liable therefore in the absence of an agreement, express or implied. This is well supported by the following decisions: *New Jersey Central Railroad Company v. MacCartney (N.J.)*, (52 Atl. 575); *Davis v. Allen (S.C.)*, (117 S.E. 547); *Railroad Company v. Evans (Mo.)*, (228 S.E. 8530).

If, however, the shipment is accepted, then the consignee, under the rule laid down in the *Fink case, supra,* implicitly agrees to pay for the freight charges, and becomes liable, as a matter of law, for the full amount of such charges, whether they are demanded at the time of delivery or not until later.

It is understood, of course, that the consignee must have a good reason for refusing the shipment, or otherwise he can be held liable by the consignor for any damages flowing therefrom.

The Time Limit for Filing Undercharge and Overcharge Claims

Section 11706(a) of P.L. 95-473 gives the carriers three years from the time the cause of action accrues (delivery or tender of delivery by the carrier) to collect all or any part of their charges. If suit in court is not instituted within that time by the carrier, action is forever barred and the liability destroyed.

When undercharges are discovered by the carrier, a balance due bill is usually submitted to consignor or consignee requesting payment. Failing to collect, and faced with the running of the statutory period, the carriers will notify the shipper or consignee that unless the additional charges are paid within a specified time, suit will be instituted. This the carrier must do since the law requires them to collect their tariff charges and there can be no extension of time beyond the three year period.

With regard to overcharges the situation is slightly different. Shippers and receivers have three years within which to begin their action. If the action is against a railroad or a water carrier, the shipper or receiver has the alternative of filing a complaint with the Commission or instituting suit in court. However, against motor carriers or freight forwarders, the statute specifies action at law only. In all cases, the following extensions of time are permitted (1) where a carrier begins civil action or collects all or part of its charges just prior to the 3 year period the time limit is extended 90 days and (2) where an overcharge claim has been filed, the time limit may be extended six months. This six month extension applies from the first declination of the last submission of the claim within the three year period.

Statute of Limitations and Transit Shipments

Some time ago, a Federal district court made an interesting decision concerning the statute of limitations issue in an undercharge suit. It held, that collection of undercharges by the railroad respondent with respect to shipments of rough hardwood lumber on which a transit rate normally would have applied was not barred by the two-year "statute of limitations" (now three years) set forth in Section 16 of the Interstate Commerce Act, although the railroad's suit for collection of the undercharges was not instituted until more than two years after the inbound shipments had been delivered to the manufacturer named as defendant in the suit, the statute of limitations being two years at that time. The Supreme Court of the United States permitted the lower court's decision to stand, as it denied a petition for certiorari in No. 738, *Arkansas Oak Flooring Co., et al, v. Louisiana & Arkansas Railway Co.*

The petition and briefs in No. 738 showed that the inbound shipments (35 carloads) of rough hardwood lumber had been delivered to the Arkansas Oak Flooring Co. at Alexandria, La., in the period from September 12, 1942, to May 12, 1943, and that these shipments had moved under a "rough material" tariff providing for a local rate of 20 cents a hundred pounds and for a credit of eight cents a hundred pounds on outbound manufactured products made from this lumber, "to the extent and in the manner specified in the tariff and within one year from the dates of the paid freight bills covering the inbound shipments of rough hardwood lumber." The tariff further specified, according to the certiorari petition, that the shipper might pay the net transit rate of 12 cents a hundred pounds on the rough hardwood lumber when it was delivered and have credit extended to it for the anticipated refund of eight cents a hundred pounds on the outbound shipments. It was stated that the L. & A. had collected the net transit rate of 12 cents when the inbound lumber was delivered, but that, "through a misunderstanding, caused or contributed to by the previous course of dealing between the parties," the Arkansas Oak Flooring Co. failed to make the outbound shipments of manufactured products which were required for the 35 carloads in question within the time prescribed in the tariff.

The railroad began its suit to recover the undercharges (the difference between the inbound transit rate of 12 cents collected and the local rate of 20 cents applicable) on September 11, 1945, according to the petition. Thus, the flooring company said, the action was not begun until more than two years after the delivery of the last of the shipments involved. It said the Federal district court in which the suit was tried, held that the term "two years" in Section 16(3)(e) of the Act meant two years from time of delivery of the manufactured products to the ultimate destination, under the transit provision of the tariff, rather than two years from the time of delivery of the shipment by the L. & A. The United States Circuit Court of Appeals for the Fifth circuit, in its decision of February

20, 1948, concluded that the tariff provision had the legal effect of postponing the accruing of the cause of action until the expiration of the period of one year allowed for shipment of the outbound products under the transit rate provision.

Statute of Limitations as to Common Carrier Other Than Railroads

Prior to June 29, 1949, there was no provision in Part II (motor carriers) of the Act similar to the statute of limitations provision of Part I (railroads) of the Act. Suits by motor carriers to collect undercharges and shippers claims for overcharges in connection with motor carriers were subject to the statute of limitations of the particular state in which the action was brought, and there is a diversity of periods of limitations so fixed by the various states.

H.R. 2759, passed by the House and referred to the Senate for action in May, 1948, proposed a two-year period of limitation within which actions might be brought for the recovery of undercharges or overcharges by or against common carriers by motor vehicle, common carriers by water, and freight forwarders. The Senate acted favorably on this bill with the result that Congress amended the Act by: adding section 204a to Part II (motor carriers); amending section 308(f) of Part III (water carriers); and, adding section 406a to Part IV (freight forwarders); the provisions of which conformed with the provisions of Part I (railroads) of the Act relating to the collection of undercharges and overcharges. These changes became effective June 29, 1949.

The Amendment of August 26, 1958, amended sections 204a, 308(f) and 406a, to provide a three-year period of limitation within which actions might be brought for the recovery of undercharges or overcharges by or against common carriers by motor vehicle, common carriers by water, and freight forwarders. This was made applicable only to causes of action accruing after August 26, 1958, the date the amendment became effective. This is identical with the limitation period provided for railroads in section 16(3).

The present statute, Public Law 95-473, provides the same statutory periods for all carriers subject to the jurisdiction of the Interstate Commerce Commisison.

Duplicate Payment of Freight Charges

For many years, claims for the recovery of duplicate payment of freight charges were handled as overcharges and as such were subject to the three year statute of limitations. However, on August 4, 1975 the Commission, in *Duplicate Payment of Freight Charges;* No. 36062 (350 I.C.C. 513) stated:

"... This report presents an issue of statutory interpretation which we have limited to a single issue—Is a duplicate payment, when defined as two or more payments for the exact amount of the tariff-published charges, an overcharge as defined by section 16(3)(g) and its related provisions?

"It is clear from a study of the relevant case law surrounding shipper-carrier payment disputes, including Commission decisions, that the issue presented has rarely been squarely faced. The vast majority of the cases cited by the parties regarding such disputes are clearly distinguishable. While it is neither necessary nor warranted to discuss the principles and findings therein, it is crucial to our conclusions here to recognize that the cases basically involved the propriety of the rate charged or its method of computation, and most were suits to recover alleged undercharges. In brief, these actions arose where the carrier, after making an improper rate calculation or application, refunded a portion of the charges already paid by the shipper, and then was forced to sue to recover the difference between the amount it then had on account and the proper tariff-published charge, which often had originally been paid by the shipper. Relevant Commission cases are, for the most part, similarly distinguishable, as they also involve questions of rate applicability. Unaware of the problem created regarding duplicate payments here at issue, we have held that any payment resulting in the carrier's receiving more than proper charges was an overcharge subject to the Act's limitations on recovery. Our interpretation here necessarily overrules these decisions.

"Under the Act, carriers have a duty to collect the tariff-published charges. Section 6(7) reads as follows:

> '*** nor shall any carrier charge or demand or collect or receive a greater or less or different compensation for such transportation of passengers or property, or for any service in connection therewith, between the points named in such tariffs than the rates, fares, and charges which are specified in the tariff filed and in effect at the time, nor shall any carrier refund or remit in any manner or by any device any portion of the rates, fares, and charges so specified, nor extend to any shipper or person any privileges or facilities in the transportation of passengers or property, except such as are specified in such tariffs.'

"The court in Vicksburg, S. & P. Ry. Co. v. Paup, (47 F. (2d) 1069 (1931)), recognized this issue and stated, 'It was the purpose of Congress, we think, to establish a uniform system regulating the dealings between shippers and carriers, whereby rebates and other discriminations should be prevented.' The duplicate payment situation bears no relation to this intent to prohibit discrimination in the rates charges or the services offered different shippers. Similarly, we do not believe the section 16 provisions, which are clearly related to this intent, should apply to the duplicate payment situation. These provisions were designed to define and limit disputes regarding the propriety of the charges assessed by carriers, and not to control disputes arising after the proper charges have been assessed and paid. We do not believe that Congress intended a situation such as this where the shipper-carrier dispute bears no relation to the conduct of the parties regarding the performance of transportation service to be subject to our jurisdiction based solely on a broad reading of certain of the Act's technical provisions.

"This interpretation, in our view logical and equitable, is not inconsistent with the statutory language. In the duplicate payment situation, the carrier has assessed and the shipper or consignee has paid the published charges. The phrase 'charges for transportation service' contained in 16(3)(g) is, in our opinion, not applicable since delivery and payment of the freight charges, proper in all respects, completes the contract for transportation services and terminates the contractual obligations of all parties thereto. We are unable to view carrier submittal to the shipper of a duplicate bill when one bill for services rendered has previously been paid as representing charges for transportation service also eliminates application of the term 'unreasonable charge' to this situation, as argued by a party herein. Further, we view the phrase 'in excess of those applicable thereto under the tariffs lawfully on file with the Commission' to have references to disputes concerning the propriety of charges assessed on any bill or adjustment thereto in relation to the transportation performed. Of course, however, it would remain true that where an inapplicable excessive amount was paid in the first instance, or after adjustment, the action to recover that amount would be based on the contract of transportation and the overcharge theory, as the dispute prevents fulfillment and termination of the contractual obligations of the parties and the carrier's responsibility under the Act.

"Our interpretation is further supported by the fact that, if duplicate payments were defined as overcharges, the cause of action pursuant to section 16(3)(e) could arise prior to the actual duplicate payment. Once it is understood that the duplicate payment bears no relation to the transportation service performed, there appears no purpose in dating accrual of the cause of action back to the date of delivery of the shipment. We do not believe Congress intended such a result.

"In reaching our conclusion we do not ignore its practical consequences. Omitting duplicate payments from the term overcharge and the relevant statute of limitations, excludes Commission consideration of these cases and places them solely within the jurisdiction of the civil courts, subject to the statute of limitations applicable to the cause of action. We do not foresee that this result will in any manner adversely affect the public interest or our pursuit of the goals of transportation regulation set forth in the Act. Carrier comment that exclusion of duplicate payments from the term overcharge and the Act's 3-year statute of limitations would result in additional unwarranted recordkeeping expense has not gone unnoticed, and we agree that there must be some point after which duplicate payment claims should not be recognized. Our decision here does not elimininate such a deadline. It does, however, change it to the applicable State statute of limitation.

"We believe our interpretation here is necessary. The carriers and their representatives are reminded that, had they acted to resolve duplicate payment claims promptly and with understanding of the reasons for which they occurred, instead of attempting to retain as a revenue source these monies paid in error and not paid for performance of transportation services, our action herein would not be necessary.

"An appropriate order is attached discontinuing this proceeding."

CHAPTER 14

CLASSIFICATION COMMITTEE PROCEDURE

General

Going back to the year 1887, we find that the early development of classification of freight by the railroads in the United States was not along any definite lines. Carriers acted independently, adopting individual classifications. It has been estimated that there were, at one time, as many as 138 distinct classifications in eastern trunk line territory, each classification built up independently of all others to serve the needs of the particular road to which it applied.

The growth of through traffic and the formation of through routes over connecting lines necessitated the establishment of classifications in addition to those adopted by each separate carrier for its own traffic. To meet this need confederations of railroad companies established classifications for through traffic in various sections of the country, for example, in eastern territory there were published: The Trunk Lines—Westbound Classification, the Eastbound Classification, the Joint Merchandise Classification, the Middle and Western States Classification, and the East and South-Bound Classification.

As a result of this multiplicity of classifications there was great confusion in the traffic situation in this respect. After the Act to Regulate Commerce took effect, unjust discriminations and other grievances, resulting from the chaotic classification condition, were among the first complaints requiring investigation. In its *Seventh Annual Report, 1893,* page 55, the Commission declared: "Our experience in making investigations and administering the law afford many illustrations of the confusion and injustice which come as the direct effect of a varying, diverse, and conflicting arrangement."

To correct the situation, the carriers formed so-called classification committees in the eastern, southern and western territories. These committees, composed of representatives of railroads operating in those territories, did not operate under the same rules of procedure. For example, the eastern committee met but twice a year to consider and act upon proposed changes, and in some cases the hearings before the classification committees were semi-public and in others private. It was not long after the establishment of these committees that the western railroads adopted a new plan by establishing a Classification Committee, whose members were not identified with any particular railroad, which was given the power to handle all classification matters and which was to be in constant session. This plan had the effect of speeding up the handling of classification subjects and eliminating the influence of individual railroads. Eventually, the eastern and southern roads reorganized their classification committees along similar lines.

Early in 1918, in view of the progress made in the matter of classification, the Interstate Commerce Commission inquired of the carriers why they could not by January 1, 1919, or sooner, effect a consolidation of the three general classifications into one volume containing one set of uniform commodity descriptions with three rating columns, one for each territory, and one set of general rules. Shortly thereafter the director of traffic of the Railroad Administration, after conference with the Commission, appointed a special committee of experienced classification men to carry out the work the Commission had in mind.

The special committee which prepared the consolidated classification consisted of the chairman of the committee on uniform classification, the chairmen of the Official, Western and Southern Classification Committees, and the Commission's classification agent. After investigation and certain recommendations by the Commission, the proposed Consolidated Freight Classification was adopted and a Consolidated Classification Committee created. The Commission's report in the *Consolidated Classification Case,* (54 I.C.C. 1), was decided July 3, 1919.

Following the Commission's report in *Docket No. 28300, Class Rate Investigation, 1939,* and *No. 28310, Consolidated Freight Classification,* (262 I.C.C. 447), May 15, 1945, the carriers organized the Uniform Classification Committee with administrative jurisdiction over the Uniform Freight Classification. This Committee consists of a chairman, vice-chairman and two members. Headquarters of the Committee is at 222 S. Riverside Plaza, Chicago, Ill. 60606. The Official, Southern and Western Classification Committees retained separate identities until September 1, 1964. Since that date, all classification proposals have been considered by the Uniform Classification Committee.

Development of Uniform Classification

In the preparation of its first docket, the Uniform Classification Committee announced that it would follow as closely as possible the Commission's admonition to produce a Classification which would "insure proper preservation of revenues for the carriers" and which would produce rates which "shall be just and reasonable as respects the shipper." Also, in compliance with admonitions of the Interstate Commerce Commission the Committee took into careful consideration all exceptions to the Classification and the proposed uniform ratings in its first docket were intended to supplant general exceptions which would otherwise destroy classification uniformity.

The admonitions from the decision of the Interstate Commerce Commission in Docket No. 28310, referred to above, are:

> "Classifications are not the correct media for remedying inequalities in territorial class rate levels, which should be accomplished by modifying the rates themselves. Where adequate justification exists for lower charges in a portion of a territory, or from and to particular points the proper remedy is the establishment of lawful exceptions to the classification, or commodity rates. The situation described does not justify exceptions that have territory-wide application and afford relatively lower rates than in the immediately adjoining or adjacent territory.
>
> "While exceptions to classifications are not in issue in this proceeding, it is difficult to understand how a particular rating on a given article or description of traffic is justified for application throughout an entire present classification territory, but is not extended to an adjoining territory. It would be unusual if the justifying conditions ended abruptly at the territorial boundaries.
>
> "We believe that the assignment to the same article of commerce of a rating which is different (in terms of percentage relation between class rates governed thereby) throughout one territory from the rating in another, because the article moves in large volume in the one as a whole and in small volume in the other, causes undue preference and prejudice between the shippers in the two territories, as well as between the traffic and between the territories. A heavy movement between particular points might be so great and regular as to justify commodity rates between such points lower than the corresponding rates under classification ratings, and heavy movements between many points in a broader, though localized, area might justify exceptions to the classification restricted to such area; but neither situation would warrant classification, or exception, ratings throughout one of the present extensive territories, different from those in the adjoining territory.
>
> " *** it is not intended that the *** findings with respect to uniformity shall prevent the making of exceptional classification ratings *** provided such exceptional ratings *** shall not tend to impair *** uniformity of Classification *** "

Docket No. 1, the first of a series of dockets to be issued and scheduled for public hearing, was issued July 15, 1947, and hearings set for seven points throughout the country in August and September.

The docket of 54 pages showed the present less-than-carload and carload ratings and minimum weights in the three classification territories and the proposed unified less-than-carload and carload rating and minimum weights on each of something over 12,000 items, grouped generally as lumber and forest products, including building woodwork; machinery and machinery parts; meats and packing house products; paper and paper articles, including paper bags and wallboard. In announcing the docket, the Committee said:

> "In the preparation of this docket, the Committee has made use of the best available information as to classification characteristics of the articles named herein. In order that the Committee may have accurate and up-to-date information, shippers are urgently requested to state the kinds of packages used and the exact dimensions, weight, and value of each package. Should there be objection to stating value per package, shippers should furnish accurate information as to weight per cubic foot and value per pound of each article as packed for shipment."

At the hearings on proposed uniform classification ratings contained in Docket No. 1, shippers were advised by the chairman of the Committee that as the hearings progressed, the Committee on Uniform Classification and the advisory committees of rail traffic officials would take all arguments presented under consideration, to judge what changes should be made in the docket proposals and after the dockets had been completed, the Committee would issue the complete proposed uniform classification in one book.

After months of hearings in various cities of the country, the Committee on Uniform Classification brought the taking of testimony to an end, on its fourth and final docket, at San Francisco, Cal., the latter part of April 1949. "The work of setting up a uniform classification is very nearly finished," Commissioner Mitchell wrote Senator Magnuson, April 12, 1950. "The single class rate scale prescribed in 1945, is now before the Commission for revision to reflect the postwar increases and this proceeding likely can be brought to the point of decision shortly after August 1. Barring unforeseen complications, the new rate and the new classification should become effective later this year (1950)." The rates and classification became effective May 30, 1952.

Applications

Representatives of shippers or carriers desiring changes in existing provisions or addition of new items in the Uniform Freight Classification are required to submit an application to the chairman of the Uniform

Classification Committee. Copies of the application form may be obtained by contacting the Committee at 222 S. Riverside Plaza, Chicago, Ill. 60606. If it is desired that the proposed change be made in the Consolidated Freight Classification, specific requests should be made on the application form.

All of the information called for on the application should be supplied. Catalogs, exhibits or cuts are often helpful to the Committee. The Committee's docket generally closes on the third Tuesday of the month, and in order to permit proper investigation and publicity, applications should be filed well in advance of the closing date. Proposals arriving after a closing date are held over to the next monthly docket.

The first section of page one of the application calls for information on the article, such as: the name of the article, full description of the article, what it is made of, and its uses.

The second section of page one calls for information as to how the article is packed or prepared for shipment; i.e., whether in bags, barrels, boxes, bundles, crates or other packages or in pieces or in bulk, when shipped L.C.L. and when shipped C.L. Also, if the article is knocked down, to what extent, or if shipped nested, to what extent. This section also calls for the dimensions of package or pieces, in inches, showing length, width and height; weight as packed for shipment, i.e., weight per package or piece in pounds and weight per cubic foot in pounds; and value per pound.

The third section of the application calls for the present classification description and Uniform, Official and Western Classification ratings, both less-than-carload and carload, the minimum carload weight and reference to the page and item of the Classification.

The fourth section of the form calls for the classification desired including the description of the article, the minimum carload weight and the less-than-carload and carload ratings desired in Uniform, Official and Western Classifications.

The second page of the application, in addition to the reasons why change or addition is requested, calls for the actual weight that can be loaded in standard cars 40 ft. 6 in. in length, 9 ft. 2 in. wide and 10 ft. high (inside measurement); volume of movement if shipped in straight carloads, territory in which moved; if shipped in mixed carloads with other articles, description of such articles and where produced.

We need not explain the reason for the Classification Committees requesting the information outlined above, as all of the information is necessary to properly determine the ratings to be assigned.

Dockets

Proposals for changes in classification rules, descriptions, ratings and minimum weights are publicized in dockets distributed by the Uniform Classification Committee. The Committee issues a regular docket every month. This docket covers all subjects other than fresh fruit or vegetable containers, which are dealt with on special dockets. In addition to the docket service of the Committee, proposals on the regular and special dockets are also listed in Part 2 of the *Traffic Bulletin*, published by Traffic Service Corporation, 815 Washington Building, Washington, D.C. 20005.

The Committee's docket service lists the proposals received, giving each proposal a subject number and indicating if it is a shipper, carrier or freight forwarder proposal. Under the subject number are shown the present classification item number, description of the article, less-carload and carload ratings and carload minimum weight. Below the present situation is shown the proposed change. A specimen of the title page and a page of subjects are shown on pages 229 and 230.

Public Hearings

Meetings of the Uniform Classification Committee, are held at Chicago, Ill., generally the second Tuesday of the month. At such meetings public hearing is afforded representatives of shippers or carriers who may wish to present statements respecting any docketed subject.

Applicants of record are notified by the Classification Committee that the Committee will receive their statements or arguments at a stated date and hour during the meeting and an appearance number is assigned indicating the time for their appearance which time may be from 20 minutes to several hours depending upon the nature of the proposed change.

Shippers presenting written arguments and exhibits at hearings are requested to furnish at least four complete copies.

Procedure By Committee

Parties at the hearing must complete an attendance slip, which requires the name, title, company affiliation, address and subject number in which the party is interested.

The chairman, at the beginning of each hearing, will announce the names of all parties in attendance and then call the person, or his representative, who filed the proposal to begin the proceedings, since he is the proponent of the subject matter. In some instances, the proponent may also introduce a second party to testify, since the issue may require the testimony of an expert witness such as a chemist or a packaging engineer.

The party testifying before the committee may present his views orally or in writing with exhibits or actual samples of the commodity or a sample of the packaged article if the proposal involves packaging. During the hearing the Committee will ask questions pertaining to the nature of the proposal and if the information cannot be furnished at that particular time, the witness may submit the information at a later date. It is not necessary that proponent be represented by counsel. However, many traffic people appearing before the Classification Committee at the public hearings are registered practioners before the Interstate Commerce Commission. All hearings before the Classification Committee are recorded on electronic tape for future reference if necessary.

Following monthly hearings on each monthly docket the Committee sits in private session and, based on the comments submitted on each subject, will review each proposal and arrive at a recommendation as to whether the proposal should be approved, disapproved or modified. In some instances either prior to formal consideration or at the public hearing, the proponent may decide to withdraw his proposal.

Following consideration on the entire docket, the Classification Committee will issue a Recommendation Advice to the Chief Traffic Officers of the railroads in the three jurisdictions—Traffic Executive Association-Eastern Railroads; Executive Committee, Southern Freight Association; and Executive Committee-Western Railroad Traffic Association—referring to the individual subjects and indicating therein whether or not the subject has been recommended, approved, disapproved or modified.

Proposals relating to changes involving ratings or minimum weights must show a justification supporting the Committee's action. Changes relating to packaging need not show the justification.

The recommendations of the Classification Committee which are submitted to the Chief Traffic Officers in the three jurisdictions are submitted in conformity with the provisions of Article III of the Interterritorial Agreement. Unless objections are expressed within fifteen days from the date of the Recommendation Advice, such recommendations, as issued by the Classification Committee, will be deemed to have the concurrence of the carriers receiving such recommendations. If a proposal is not objected to, the recommended change will be published in the next available supplement issued to the Classification in effect at that time. However, a carrier objection registered in connection with a proposal stays the recommendation issued by the Classification Committee and the matter is remanded back to the Chief Traffic Officers of the railroads in the particular jurisdiction where the objection was registered. Upon review of the recommendation objected to the objection may be re-affirmed by the entire jurisdiction, or the recommendation issued by the Classification Committee may be affirmed. In either case the action taken finalizes the handling of the proposal.

Announcement of Disposition of Docketed Subjects

Where changes are made as the result of petitions, advice is given to those who may have appeared at hearings in connection with such subjects or from the records are shown to be entitled to advices. Where petitions are declined, advice is given to those of record concerned in the subject. These notices, however, will not disclose all the facts in possession of the Committee which leads to such conclusions. It may be added, however, that the Committee is at all times available and is willing to explain the controlling causes which seem to justify the declination of any petition.

Publication of Changes

After the Uniform Classification Committee has reached a decision, advice of the action taken on each subject is issued to the participating carriers. In the case of proposals which have been approved as proposed or on a modified basis, if no objections are received from participating carriers, a supplement to the classification is issued embodying the changes made. The determination of the effective date is based on the legal requirement of thirty days notice. However, forty-five days between the date of issue and date effective is usually allowed for filing and posting.

UNIFORM CLASSIFICATION COMMITTEE

ROOM 1106, 222 S. RIVERSIDE PLAZA, CHICAGO, ILLINOIS 60606

TELEPHONE: 648-7944 & 7949 AREA CODE 312

J. D. SHERSON
CHAIRMAN

G. F. EARL
VICE-CHAIRMAN

F. W. PARILLA
S. D. KENNEDY
MEMBERS
R. J. NOVAK
ASST. TO CHAIRMAN

J. P. BEITZEL
R. M. WALSH
SPECIAL REPRESENTATIVES
CONTAINER SECTION

IN REPLY REFER TO FILE

DOCKET 3__

JANUARY , 19

SUBJECTS OTHER THAN FRESH FRUITS OR

VEGETABLE CONTAINERS

THE FOLLOWING SUBJECTS ARE HEREBY DOCKETED FOR CONSIDERATION. PUBLIC HEARING WILL COMMENCE WITH JANUARY 11, 19 .

PROPOSED CHANGES IN CLASSIFICATION PROVISIONS AS SHOWN IN THIS DOCKET, WHETHER INITIATED BY SHIPPERS, CARRIERS OR COMMITTEE, AFTER PUBLIC HEARING, MAY BE DEVIATED OR VARIED FROM AS WARRANTED AFTER CONSIDERATION OF THE INFORMATION AND FACTS SUBMITTED.

HEARINGS BY APPOINTMENT. IF NO HEARING IS REQUESTED OR ASSIGNED, PROPOSALS WILL BE CONSIDERED BASED SOLELY UPON AVAILABLE INFORMATION. FOR HEARING TIME ASSIGNMENT APPLY TO UNIFORM CLASSIFICATION COMMITTEE.

ISSUED DECEMBER 23, 19

UCC LISTS 21 & 25 - 635-W

DOCKET 3__

(S) SUBJECT 2

PRESENT

Item 11945, UFC :

	LCL		CL
Boiler cleansing, preserving, scale preventing or scale removing compounds, dry, in barrels or boxes or in metal cans in crates; also CL, in Package 796	55	36,000	30

PROPOSED

Amend Item 11945, UFC , by the addition of the underscored as shown below:

	LCL		CL
Boiler cleansing, preserving, scale preventing or scale removing compounds, dry, in bags, barrels or boxes or in metal cans in crates; also CL, in Package 796	55	36,000	30

(C) SUBJECT 3

PRESENT

Item 77170, UFC :

	LCL		CL
Peony buds, fresh cut, loose or in packages	...	14,000R	100

PROPOSED

Cancel . . . . Account obsolete.

(C) SUBJECT 4

PRESENT

Item 88100, UFC :

	LCL		CL
Stolons (chopped grass) in bags	200	14,000R	85

PROPOSED

Cancel . . . . Account obsolete.

National Classification Board

of the Motor Carrier Industry

1616 P STREET, N.W. WASHINGTON, D. C. 20036 202-797-5308

PROPOSAL FOR CHANGE IN THE NATIONAL MOTOR FREIGHT CLASSIFICATION

DATE: ____________________

ONE COPY OF PROPOSAL AND SUPPORTING EXHIBITS OR STATEMENTS **MUST** BE SIGNED AND SUBMITTED BY THE PROPONENT. ALL QUESTIONS MUST BE ANSWERED AS COMPLETELY AS POSSIBLE. WHEN ACTUAL DATA IS NOT AVAILABLE BECAUSE PRODUCT IS NEW, PLEASE ESTIMATE. FAILURE TO COMPLETE THE PROPOSAL FORM MAY RESULT IN DELAY, AS THE PROPOSAL WILL NOT BE PROCESSED FOR DOCKETING UNTIL PERTINENT INFORMATION IN THIS FORM HAS BEEN FURNISHED. ALL INFORMATION WILL BE TREATED IN A CONFIDENTIAL MANNER, UNLESS OTHERWISE AUTHORIZED UNDER 2 (a) and (b).

PLEASE TYPE OR PRINT CLEARLY

1) PROPONENT'S COMPANY NAME ____________________

STREET ADDRESS ____________________

CITY __________ STATE __________ ZIP __________

INDIVIDUALS' NAME __________ TITLE __________

SIGNATURE __________ TELEPHONE NO. __________

2) **PROPONENT IS REQUIRED TO ANSWER THE FOLLOWING QUESTIONS:**

(a) May your name be quoted as the proponent of the proposal? Yes _____ No _____

(b) May the information contained in this proposal be released to the following interested parties? Carriers _____ Shippers _____ Others _____

(c) What supporting data is attached? Photos _____ Blueprints _____ Catalogs _____ Samples _____ Advertising _____ Labels _____
Empty Retail Sales Containers _____

3) Description of Article(s): ____________________

4) Trade nomenclature or registered trade name: ____________________

5) Sales invoice nomenclature: ____________________

6) PRESENT CLASSIFICATION: (Show specific NMFC item number and description under which commodity is now being classified)

ITEM NO.	DESCRIPTION	CLASSES LTL	CLASSES TL	MINIMUM WT. FACTOR

PROPOSED CLASSIFICATION: (Show description, classes and minimum wt. factor exactly as you propose them to be established in the Classification. IF a packaging proposal amend item accordingly)

ITEM NO.	DESCRIPTION	CLASSES		MINIMUM WT. FACTOR
		LTL	TL	

7) SHIPPING CHARACTERISTICS AND VALUE:

TYPE OF PACKAGE USED (1)	MODEL OR STYLE NO.	NO. OF ARTICLES PER PACKAGE	OUTSIDE DIMENSIONS OF SHIPPING PACKAGE (2)			WEIGHT AS PACKED FOR SHIPMENT (POUNDS)		WHOLESALE (CLAIM) VALUE PER PACKAGE
			Length	Width	Height	Per Package	Per Cu. Ft.(3)	

(1) Wooden or Fibreboard Box; Crate; Drum (and capacity in gallons); Wrapped or Non-wrapped Bundle; or Roll (paper, plastic or cloth wrapped); Paper, Plastic or Cloth Bag; or Other Type of Package.

(2) If package is curved or irregular shape use extreme outside measurements.

(3) To ascertain the "weight per cubic foot" multiply the three dimensions of the article as packed for shipment and where the result is in cubic inches, divide by 1728 to reduce to cubic feet, then divide the weight by the number of cubic feet thus ascertained. To obtain the cubic feet of space occupied by a drum, pail, etc., square the greatest diameter and multiply by the height.

8) COMMODITY CHARACTERISTICS:

(a) What are materials from which commodity is made, by percentage of volume and weight? ____________________

____

(b) Is commodity subject to U.S. Department of Transportation Regulations Governing Explosives and Other Dangerous Articles? Yes ______ No ______

If yes, What commodity description and label are required? ____________________

____

(c) Is commodity Liquid ______, Paste ______, Dry ______, when shipped?

(d) Does commodity require Heat ______ Refrigeration ______ Temperature Control ______ while being transported?

(e) If commodity is wooden, is it "in the rough" ______, "in the white" ______, or "finished" ______?

"IN THE ROUGH" REFERS TO WOODEN ARTICLES THAT ARE NOT FURTHER MANUFACTURED THAN SAWED, HEWN, PLANED OR BENT.

"IN THE WHITE" REFERS TO WOODEN ARTICLES THAT ARE FURTHER MANUFACTURED THAN "IN THE ROUGH", BUT INCLUDING NOT MORE THAN ONE COAT OF PRIMER.

"FINISHED" REFERS TO WOODEN ARTICLES AFTER THEY HAVE PASSED THE STATE OF "IN THE WHITE".

(f) If commodity is not wooden and is considered to be unfinished, what processes have been performed and what processes remain to be performed before consumption or use? ____________________

____

____

____

9) FORM OF SHIPMENT:

(a) Is commodity SU______, SU in sections ______.
"SU" (Set Up) refers to articles in their assembled condition, or disassembled, folded or telescoped but not meeting the conditions described in paragraphs (b) through (f) below. Where an article does meet the conditions described in paragraphs (b) through (f), but such provisions are not published in item descriptions, the SU classes will apply.
"SU in sections" or "In SU sections" refers to articles taken apart in sections.

(b) Is commodity "KD" ______, KD flat______?
"KD" (Knocked Down) refers to an article taken apart, folded or telescoped in such a manner as to reduce its bulk at least 33 1/3 percent from its normal shipping cubage when set up or assembled.
"KD flat" refers to an article taken apart, folded or telescoped in such a manner as to reduce its bulk at least 66 2/3 percent from its normal shipping cubage when set up or assembled.

(c) Is commodity folded ______, folded flat ______?
"Folded" refers to an article folded in such a manner as to reduce its bulk at least 33 1/3 percent from its normal shipping cubage when not folded.
"Folded flat" refers to an article folded in such a manner as to reduce its bulk at least 66 2/3 percent from its normal shipping cubage when not folded.

(d) Is commodity nested ______, nested solid ______?
"Nested" refers to three or more different sizes of an article placed each smaller within the next larger or three or more of the same article placed one within the other so that each upper article will not project above the next lower article by more than one-third of its height.
"Nested solid" refers to three or more of the same article placed one within the other or upon the other so that the outer side surfaces of the one above will be in contact with the inner side surfaces of the one below and so that each upper article will not project above the next lower article more than ¼ inch.

(e) Is commodity compressed______, if so, by machine______, by hand______?

(f) Is commodity disassembled ______, partially______, completely______?

(g) Is commodity wheeled, if so is it shipped with wheels on ______, off______?

10) CHARACTERISTICS OF SIZE AND METHODS OF SHIPMENT:

(a) What are predominant areas of movement? Nationwide______, East______, South ______, Midwest______, Southwest______, Farwest______.

(b) What are predominant methods and sizes of shipments?
(1) Less than truckload lots ______%, average size of shipment ______ lbs.
(2) Straight truckload lots ______%, average size of shipment ______ lbs.
(3) Mixed truckload lots ______%, average size of shipment ______ lbs.

(c) What are the usual terms of shipment?
(1) Freight prepaid by shipper ______.
(2) Freight collect______.
(3) Collect on delivery (C.O.D.) ______.
(4) Freight prepaid by shipper and added to customer invoice ______.

(d) What is annual tonnage of this commodity by motor common carrier? ______

(e) What percentage of movement is by motor common carrier? ______%

(f) What percentage of movement is subject to class rates? ______%

11) CLAIMS EXPERIENCE (Motor Common Carrier Shipments):

(a) Claims filed by: Shipper ______, Consignee ______
(b) Number of claims (one year period) ______, What period? ______ through ______
(c) Dollar amount of claims ______ Amount Paid ______

(d) Total number of shipments made (one year period) ______
(e) Total amount of freight paid (one year period) ______

12) What are major competitive commodities? ______

13) (A) Names and addresses of Manufacturers or Shippers (when proponent is carrier)

(B) Name and address of Competitors (when proponent is manufacturer)

14) Names of originating carriers (when proponent is shipper or manufacturer)

15) STATE FULL REASON(S) FOR PROPOSAL CHANGE OR ADDITION. (If, as justification, proponent refers to ratings of other articles in the Classification, complete information as to weight, value, loadibility and other pertinent factors of such comparable items, should be given. Use separate sheet if additional space is needed.

VOL. 134 NO. 26 PART III DECEMBER 29, 1979

TRAFFIC BULLETIN

NATIONAL CLASSIFICATION BOARD

1616 'P' STREET, N.W.
WASHINGTON, D.C. 20036

DOCKET NO. 801

JANUARY, 1980

For consideration of proposals for change in National Motor Freight Classification rules, descriptions, ratings, packaging and minimum weights as set forth herein. Hearings will be held on individual subjects only by appointment. Hearings will begin at 1616 'P' Street, N.W., Washington, D.C. 20036, on January 22, 1980.

Request for Hearing appointments (via Speaker-Phone or personal appearance) should be made to the Secretary (202-797-5308) promptly for allotment of time for presentation of facts in support of or in rebuttal to proposals. Ten complete copies of written arguments and exhibits should be furnished.

Proposed changes in this docket will be decided after public hearings, upon consideration of available facts. Effort is made to avoid publication of proposals which are obviously without merit or impossible of adoption. Publication of proposals on this docket does not imply endorsement by the National Classification Board.

The Classification Board is authorized to deviate from proposed changes and will accord consideration to all factors involved, including descriptions, ratings, rules, minimum weights and packaging requirements.

Explanation of letters: **(C)** means proposed by a carrier; **(S)** means proposed by a manufacturer or shipper; **(B)** means change proposed by the National Classification Board.

Use of the symbol ▶ in Item descriptions, classes or minimum weight factors indicates change.
Use of the symbol **O** in Item descriptions indicates commoditiy or commodities may be subject to special federal regulations concerning the shipping of hazardous materials.

DOCKET	CLOSING	MAILING	HEARINGS BEGIN
801, January, 1980	December 7, 1979	December 28, 1979	January 22, 1980
802, March, 1980	February 1, 1980	February 22, 1980	March 18, 1980
803, April, 1980	March 7, 1980	March 28, 1980	April 22, 1980
804, May, 1980	April 4, 1980	April 25, 1980	May 20, 1980
805, July, 1980	May 30, 1980	June 27, 1980	July 22, 1980
806, September, 1980	August 1, 1980	August 22, 1980	September 16, 1980
807, October, 1980	September 5, 1980	September 26, 1980	October 21, 1980
808, November, 1980	October 10, 1980	October 31, 1980	November 25, 1980

NATIONAL CLASSIFICATION BOARD

Frank T. Grice, Chairman
Gordon H. Anderson
George M. Beck
Robert F. Bobowski
Peter M. Lafakis
Robert S. Leonard
Jane E. McIntyre
William J. Schweikert
Jerry D. Stone

Fred C. Merriwether, Secretary

ISSUED: JANUARY 22, 1980

NATIONAL CLASSIFICATION BOARD DOCKET 801

Item	Description	LTL	TL	MW
	SUBJECT NO. 8 (S)			
	PRESENT			
	PADS OR PADDING:			
149220	**Pads or Padding,** air filtering or evaporative cooler, see Note, item 149222, expanded paper or expanded paper and expanded aluminum foil, framed or not framed, in boxes	300	300	AQ
149222	NOTE—The term 'air filtering pads or padding' is embracive of filtering pads or padding such as are used in air conditioners or paint spraying booths.			
	PROPOSED			
149220	**Pads or Padding,** air filtering or evaporative cooler, see Note, item 149222, expanded paper or expanded paper and expanded aluminum foil, framed or not framed, in boxes, ▶having a density of:			
▶Sub 1	Less than 2 pounds per cubic foot	300	300	AQ
▶Sub 2	2 pounds or greater per cubic foot	250	125	10
149222	NOTE—No Change.			
	JUSTIFICATION: New sub and classes being added to provide for these products when the density is two pounds per cubic foot or greater.			
	SUBJECT NO. 9 (B)			
	PRESENT			
155710	**Pipe Covering,** other than pipe covering shapes, viz.: Asbestos or cloth, impregnated or coated with asphalt, pitch, plastic, tar or wax, with or without plastic film backing or separator sheet; Asphalt, pitch or tar composition in sheet or tape form, with or without plastic film backing or separator sheet; Plastic film, with adhesive backing, with or without plastic film separator sheet; In boxes or in wrapped bundles or rolls	60	35	36
155720	**Pipe Joint Covering or Pipe Joint Covering or Coupling Sealing Molds,** paperboard or pulpboard combined with felt, with steel installation straps in same or separate boxes:			
Sub 1	Curved, exceeding 3/16 of a circle	300	300	AQ
Sub 2	Not curved, or curved to not exceeding 3/16 of a circle	125	77½	14
	PROPOSED			
155710	**Pipe Covering, etc.** ▶Cancel; see item ▶A			
155720	**Pipe Joint Covering,** etc. ▶Cancel; see item ▶B			
	Concurrently add the following new items:			
▶A NEW	**Covering,** pipe, other than pipe covering shapes, viz: Asbestos or cloth, impregnated or coated with asphalt, pitch, plastic, tar or wax, with or without plastic film backing or separator sheet: Asphalt, pitch or tar composition in sheet or tapr form, with or without plastic film backing or separator sheet: Plastic film, with adhesive backing, with or without plastic film separator sheet: In boxes or in wrapped bundles or rolls	60	35	36
▶B NEW	**Covering,** pipe joint, **or Pipe Joint Covering or Coupling Sealing Molds,** paperboard or pulpboard combined with felt, with steel installation straps in same or separate boxes:			
Sub 1	Curved, exceeding 3/16 of a circle	300	300	AQ
Sub 2	Not Curved, or curved not exceeding 3/16 of a circle	125	77½	14
	JUSTIFICATION: Reindexing by noun is in the interest of tariff simplification and clarification.			

NATIONAL CLASSIFICATION BOARD
AMERICAN TRUCKING ASSOCIATIONS, INC.

RULES OF PROCEDURE FOR CHANGES IN THE NATIONAL MOTOR FREIGHT CLASSIFICATION

Rule 1. National Classification Board—Composition and Functions

"The National Classification Board (hereinafter called the Board) shall be composed of not less than three and not more than seven full-time employees, one of whom shall be the Chairman. It shall consider proposals for changes in the descriptions of articles, minimum weights, packing requirements, ratings, rules or regulations in the National Motor Freight Classification (hereinafter called the Classification), hold public hearings, recommend the disposition of proposals and instruct the Publishing Agent of the Classification as to resulting changes in the Classification.

Rule 2. Proposals

"All proposals for changes in the descriptions of articles, minimum weights, packing requirements, ratings, rules or regulations in the Classification, shall be submitted in writing to the Board. Proposals may be filed by any person, firm or corporation having an interest in the contents of the Classification.

Rule 3. Proposal Forms

"The Board shall provide a suitable form for the submission of proposals, and it may require the use of such form fully executed, in duplicate by the proponent.

Rule 4. Docketing

"Proposals received by the Board which comply with the provisions of Rule 3 will be placed on the first available docket for hearing before the Board.

Rule 5. Changes without Docketing

"Changes in the Classification made necessary by law or by order of a regulatory body or for clarification, may, at the discretion of the Board, be handled by the Board without public notice or observance of regular docket procedure. However, in such cases, the Chairman of the Board shall send a bulletin to all members of the National Classification Committee (herein called the Committee) notifying them of such action.

Rule 6. Public Hearings

"Public hearings shall be conducted on all pending proposals at such intervals, and at such places as the board may deem proper and adequate. However, the Board shall hold no less than three hearings during any calendar year, unless fewer than five proposals are pending in which event the Board may defer such hearing.

Rule 7. Disposition of Proposals

"A majority vote of the members of the Board shall govern its action. If the Board reports a proposal as adopted and no objections as provided in Rule 9 herein are received, the Board shall instruct the Publishing Agent to proceed with the publication thereof.

"In disposing of a docket subject, the Board may, in its discretion, deviate from the terms of a proposal, but may not unduly broaden the issues defined in the published docket, giving due consideration to the evidence and research made by the Board.

Rule 8. Reconsideration

"The Board may, in its discretion and for good cause shown, reconsider any docket subject upon which a disposition notice has been issued; provided, that reconsideration shall not be granted after the effective date of publication in the Classification. When the Board reaches its conclusions upon reconsideration, a revised disposition notice subject to all of the procedural rules applicable to initial disposition including the appeal procedure, shall be issued. The Board may not reopen for reconsideration any docket subject previously appealed and acted upon by the Committee under the established appeal procedure.

Rule 9. Appeals

"Upon receipt within thirty days after date of notice of disposition by the Board of written objections with reasons therefore from fifteen participating carriers in the Classification or from eight members of the Committee, the decision of the Board shall be automatically appealed to the Committee through its Secretary.

"If the Board reports a proposal as having failed of adoption, the proponent thereof may appeal the decision to the Committee within thirty days after date of notice of disposition by the Board by written notification to the Secretary of the Committee setting forth the grounds for appeal.

Rule 10. Hearings on Appeals

"The full Committee shall hear all appeals. Any person interested in a proposal which is the subject of a Board disposition that has been appealed shall, upon written request, be allotted such time as may be considered proper and reasonable for an appearance before the Committee.

Rule 11. Disposition of Appeals

"A majority of the Committee present shall decide whether the recommendation of the Board appealed from shall be approved, disapproved or submitted to the Board for further consideration.

Rule 12. Notice and Publication

"Proponents of proposals, all other parties indicating an interest therein, and all members of the Committee, shall be given notice by U.S. Mail of hearings thereon by the Board (or by the Committee on appeal.) Like notice shall be given to the public by publication in Transport Topics, not less than fourteen (14) days preceding such hearings, and through any other medium which the Committee may consider necessary or appropriate.

"Notice of Board Dispositions, appeals to the Committee and Committee decisions shall be given to the parties and members of the Committee by U.S. Mail. Notice of such dispositions, appeals and decisions shall be published as provided above.

Rule 13. Independent Action.

"Every party to this Agreement shall have the free and unrestrained right to take independent action at any time either before or after any determination arrived at through any procedure provided in these rules.

"Nothing in these rules shall prevent any participating carrier from publishing or causing to be published or concurring in any Classification or in any tariff or tariffs containing exceptions to the Classification or to any or all provisions thereof.

"Should any party to this Agreement give written notice of independent action as to any proposed change in the Classification at any time within 30 days after the date of notice of disposition has been issued, the publishing agent will withhold publication of the proposed change for a period of 30 days from such notice of independent action so that such party may cause independent publication in accord with the said independent action.

Rule 14. Amendments

"These rules may be amended to any regular or special meeting of the Committee by a two-thirds majority vote, a quorum being present and voting, provided that thirty (30) days' notice in writing of the proposed amendments to the rules shall have been given by the United States mail to all members of the Committee."

CHAPTER 15

CARRIER RATE COMMITTEE PROCEDURE

General

Rate committees have attained such a prominent place in rate making and rate adjustments that it is necessary to outline some of their more important activities.

Among the first rate committees established after enactment of the Act to Regulate Commerce was the Trans-Missouri Freight Association formed in 1889 and the Joint Traffic Association formed by the eastern railroads. With the passage of the Sherman Anti-Trust Act in 1890, these associations were subject to attack by the Attorney General for violation of the anti-pooling section of the commerce act and of the anti-trust law. In the *Joint Traffic Association Case (United States v. Joint Traffic Association,* (76 Fed. Rep., 895)), the decision of the circuit court dismissing the petition was affirmed by the circuit court of appeals at the conclusion of the argument on March 19, 1897. On the Monday following (March 22, 1897) the United States Supreme Court handed down its decision in the case of *United States v. Trans-Missouri Freight Association,* (166 U.S. 290), to the effect that the agreement constituting that railway association was in conflict with the anti-trust law.

Notwithstanding the sweeping ruling of the Supreme Court in the Trans-Missouri Case—that any agreement between competing railroad companies for the maintenance of rates is unlawful—a period of one year and seven months elapsed after that decision before the Joint Traffic Association agreement, which expressly declared that its purpose was the maintenance of rates, was held by the Supreme Court—*United States v. Joint Traffic Association,* (171 U.S. 505)—to be illegal.

It was not long after these decisions that the Supreme Court recognized that the Anti-Trust Act involved only unreasonable agreements in restraint of trade, which became the accepted judgment of the Court and was known as "The Rule of Reason." In this connection, the Court in *United States v. Union Pac. R. Co.,* (226 U.S. 61), and *United States v. Reading Co.,* (226 U.S. 324), stated:

> "Applying the rule of reason to the construction of the Sherman Act, it was held in the Standard Oil Case, *supra,* distinguishing the Trans-Missouri and Joint Traffic Cases, *supra,* that, as the words 'restraint of

trade' at common law and in the law of this country at the time of the adoption of the antitrust act only embraced acts or contracts or agreements or combinations which operated to the prejudice of the public interests by unduly restricting competition, or unduly obstructing the due course of trade, or which, either because of their inherent nature or effect, or because of the evident purpose of the acts, etc., injuriously restrained trade, that the words as used in the statute were designed to have, and did have a like significance. The statute did not forbid or restrain the power to make normal and usual contracts to further trade by resorting to all normal methods, whether by agreement or otherwise, to accomplish such purpose. Not that acts which the statute prohibited could be removed from the control of its prohibitions, by a finding that they were reasonable, but the duty to interpret which inevitably arose from the general character of the term 'restraint of trade' required that the words 'restraint of trade' should be given a meaning which would not destroy the individual right to contract, and render difficult, if not impossible, any movement of trade in the channels of interstate commerce—the free movement of which it was the purpose of the statute to protect."

In the first half of 1910, numerous carriers gave notice of general advance in rates, and it was commonly understood that other carriers would shortly take similar action. There were numerous objections to these changes and the Attorney General of the United States brought suit alleging violation of the Sherman Anti-trust Act. As a result of conferences between the Government, authorities and representatives of the carriers, the dates on which the proposed advances were to become effective were postponed pending the passage of the *Mann-Elkins Act,* 1910, which granted the Interstate Commerce Commission the power of suspension, enabling it to institute investigations thereunder.

With the many changes in the Commerce Act, the carriers more and more were required to confer with each other and reach agreements of one kind or another. Also the Commission's investigations and general territorial rate decisions made joint action on the part of the carriers absolutely necessary to the proper conduct of hearings and the working out of the rates resulting from these decisions.

During World War I, all railroads of importance within the United States were placed under Government control in charge of a director general. The organization of the railroads under the director general included a number of regional freight traffic committees that considered traffic problems and rates in allotted territories. These were made up not only of railroad traffic men, but shippers were represented on each. It was believed that the shippers were given representation for the reason that certain provisions of the Federal Control Act made the director general the sole judge of the propriety of newly made rates and took from the public the remedy of seeking and obtaining suspension of proposed rates by the Interstate Commerce Commission. It was felt that, since a rate once filed under those circumstances was not subject to suspension, shippers ought, in fairness, to have some voice in the making of the rate before filing.

When the railroads were returning to private ownership, on March 1, 1920, the rate committees, considerably modified, were retained. Their value had been demonstrated during the Federal control period. The carriers reorganized these committees on a territorial basis and eliminated therefrom shippers' representation. The right of suspension, of course, again became operative with the cessation of federal control.

The rate committee, in 1920, became an essential part of the carrier's and shipper's program for the establishment of orderly ratemaking procedure. The years have proven its value both to shipper and carrier.

It is the consensus that the interests of both shippers and carriers are best served through the medium of the rate committees, because the committee provides: (a) a place through which shippers and carriers may file rate proposals in which other shippers and carriers may have a direct or indirect interest; (b) a place from which proposals can be distributed and publicized so that all interested parties may be advised at all times of proposed changes; (c) the machinery by means of which rate changes can be analyzed from both the shippers and carriers standpoint; (d) a place where shippers and carriers can meet and discuss their rate problems; (e) the machinery for announcing approved changes in a uniform manner; and (f) a tariff bureau which publishes the agency tariffs for account of member roads, thus effecting great economy in the expense of compilation and printing.

Anti-trust Suits

In 1944, there were two important developments of significance in connection with the application of the Sherman Anti-trust Act to transportation agencies. The first of these was the acquittal by a grand jury in the federal district court, on April 6, of the *Middle-west Motor Freight Bureau* of Kansas City, Mo., and a number of other similar organizations as well as motor carriers and individuals who had been indicted for conspiracy to violate the antitrust laws. The second development was the filing of a complaint by the federal government in federal district court at Lincoln, Nebraska, charging violations of Sections 1 and 2 of the Sherman Act by the Association of American Railroads and certain railroad companies and individuals, and seeking the dissolution of the Association of American Railroads.

In June of 1944, the State of Georgia filed a petition in the Supreme Court of the United States for leave to file an original bill of complaint in that court against various railroads attacking the rate procedures of the Southern and Eastern railroads and contending among other things, that they were in violation of the Anti-trust Act. The Supreme Court, by a five to four decision, decided to permit the case to proceed as an original proceeding in the Court.

The actions of the Department of Justice in the Lincoln case and the Southern Governors in the Georgia case brought a realization among railroads and shipping interests that legislation was needed to permit carrier rate-making processes to have immunity from the Anti-trust Act.

In discussing this situation in its *Fifty-Eighth Annual Report, 1944,* the Interstate Commerce Commission said:

> "Until about five years ago, the view was rather generally held that for all wrongs for which the parties are afforded a remedy by terms of the interstate commerce act, to that extent the interstate commerce act supersedes the anti-trust laws.
>
> "While such a view perhaps requires qualification in the light of more recent court decisions, it is supported to some extent by the evident disposition on the part of Congress to negative the application of the anti-trust laws to certain transactions of carriers subjected to regulation under the interstate commerce act, section 5(11) of which affords an example. Also it is significant that the statement of national transportation policy added to that act in 1940 contained the following:
>
> > " ' . . . to encourage the establishment and maintenance of reasonable charges for transportation services, without unjust discrimination, undue preference or advantage, or unfair or destructive competitive practices . . . .'
>
> "In our opinion, there is danger that undue breadth in interpreting and applying the Sherman Act would interfere with carrying out the stated policy against 'unfair or destructive competitive practices.' We believe therefore that there is just as much present need for legislation to prevent any such outcome and to clarify the rights and duties of carriers subject to the interstate commerce act as there was at the time of the first session of the 78th Congress when two bills which would have that effect were introduced: S.942, by Senator Wheeler, 'To amend the interstate commerce act to provide for agreement between common carriers by railroad, between common carriers by pipeline, between common carriers by motor vehicle, between common carriers by water and between freight forwarders, for the making and filing of rates, fares, charges and classifica-

tions for transportation of passengers and property, and for other purposes,' and H.R. 2720, by Congressman Bulwinkle, 'To amend section 1(4) of the interstate commerce act . . .'

"Although neither of these bills, in our opinion, dealt adequately with the problem, we were entirely in sympathy with their broad objective."

Congressional Action

Both S. 942 and H.R. 2720 "died" with the adjournment *sine die* of the Seventy-Eighth Congress. These bills were reintroduced January 3, 1945, when the Seventy-Ninth Congress came into being. They were assigned new numbers—S. 337 (Reed) and H.R. 167 (Bulwinkle)—and referred to appropriate committees for consideration. In March 1945, Representative Bulwinkle introduced H.R. 2536, as a substitute for H.R. 167. After extensive hearings, suggestions and recommendations by shippers, railroads, associations and the Interstate Commerce Commission, in behalf of the passage of the bill, the measure was "killed" in the Senate.

Again, with the opening of the 80th Congress, "the Bulwinkle bill" was introduced and given the numbers H.R. 221 and S. 110. Hearings were immediately set and again the "Reed-Bulwinkle bill" received much support from the public at-large, including the Interstate Commerce Commission. In July, 1947, H.R. 221 was ordered reported by the House committee on interstate and foreign commerce, but House action failed to materialize before adjournment.

After reviewing the support expressed for the legislation in the committee hearings and noting the interests favoring the bill, the majority report set forth a statement of the need for the legislation. It continued with a discussion of "precedents for legislation of the character here proposed" and a discussion of the necessity of joint action on the part of the carriers.

In its introductory paragraphs, the majority report observed that H.R. 221 would amend the Interstate Commerce Act by adding after section 5 thereof a new section 5a.

"Under the proposed section 5a," it said, "(1) common carriers (including freight forwarders) subject to the interstate commerce act are authorized to apply to the Interstate Commerce Commission for approval of agreements between such carriers; (2) the Commission is authorized to approve such agreements (except certain types of agreements the approval of which is prohibited) when the Commission finds that, by reason of furtherance of the national transportation policy declared in the interstate commerce act, there should be relief from the anti-trust laws with respect to the making and carrying out of such agreements; and (3) anti-trust law relief is granted (but no relief is granted from any prohibition, duty, or obligation under the interstate commerce act) with respect to the making and the carrying out, subject to terms and conditions prescribed by the Commission, of agreements so approved by the Commission."

The committees' discussion of the need for the legislation included the following:

"It is recognized by all who are familiar with the problems of transportation that the carriers subject to the interstate commerce act cannot satisfactorily meet their duties and responsibilities thereunder, and the basic purposes of that act cannot be effectively carried out, unless such carriers are permitted to engage in joint activities to a substantial extent. The public interest will not be served if there is permitted to continue the existing state of uncertainty as to the extent to which carriers may engage in joint activity without risk of violating the anti-trust laws."

On May 28, 1948, the Senate adopted by a voice vote the Bulwinkle bill to exempt Interstate Commerce Commission-approved agreements of regulated carriers as to rates and charges from application of the anti-trust laws. The action of the Senate, amounting to approval of the bill as amended by the House, completed action by Congress on the measure and cleared it for action by the President.

On June 12, President Truman vetoed the bill and on June 17, Congress overrode the President's veto, so the Reed-Bulwinkle Bill became section 5a of the Interstate Commerce Act. Following this the Georgia case was dismissed on motion of the parties.

Interstate Commerce Commission's Order in Connection with Section 5a

On July 6, 1948, the Interstate Commerce Commission issued an order, "Rules and Regulations Governing Applications Under Section 5a of the Interstate Commerce Act, as Amended, for Authority to Establish or Continue Agreements Between or Among Carriers."

This order prescribed the rules and regulations to be followed by carriers when filing agreements under Section 5a and also the form and contents of applications filed with the Commission, the exhibits required, and the procedure to govern the execution, filing and disposition of an application.

Aside from furnishing 20 copies of the original application and supporting papers, it is required that a copy of the application be served on the regulatory body having jurisdiction as to rates, fares, or charges of each state, territory or district embraced in the scope of the agreement. A public notice is issued by the Commission and filed with the director of the Federal Register when an application is filed, and the notice indicates how a hearing on the application may be obtained.

The first agreement approved by the Commission under Section 5a was the application of the Western Traffic Association on behalf of the western railroads which became effective December 15, 1949. Since that time, all other major rail, motor, water and freight forwarder associations, committees and bureaus have filed applications for relief from the anti-trust laws under Section 5a. Most of these have been approved subject to certain modifications in terms and conditions. The most prominent of these are that every carrier party to the agreement must have the free and unrestrained right of independent action, and no carrier's membership may be cancelled, except for non-payment of dues.

Some indication of the Commission's feeling with regard to the purpose and functions of rate bureaus may be obtained from the following excerpts from two of their decisions:

> "The congressional purpose of Section 5a is to provide a means whereby parties to an agreement relating to procedures for joint consideration and establishment of rates, charges, etc., may be relieved from operation of anti-trust laws and thereby protect the public and national interest in more stable conditions in transportation, when such agreement has been approved by the Commission. *Ahnapee & W. Ry. v. Akron & B.B. R.R.*, (302 I.C.C. 265).
>
> "Systematic procedures are essential to proper functioning of rate bureaus in the interests of shippers, localities, and carriers. The interrelated nature of the rate structure, competitive conditions which brought this about, preponderance of interline over single line movement of freight, the network nature of the railroad system which permits the free movement of commerce, contributed to organization of rate bureaus for joint consideration of rate matters. *Illinois Freight Ass'n. Agreement,* (283 I.C.C. 17)."

After a lengthy investigation into the organization and procedures of carrier rate-making associations, the Commission, on January 23, 1976, issued its report in Rate Bureau Investigation; Ex Parte No. 297 (351 I.C.C. 437). Their findings were as follows:

> "1. Rate bureaus assist the making of appropriate rates.
>
> "2. Procedural changes would foster actions more favorable to bureau members, shippers, and the general public; such changes being contained in the body of the report.
>
> "3. The right of independent action does not adversely affect the rate structure.

"4. A system need not be established by the Commission to monitor public hearings before rate bureaus, but that formal minutes, and not verbatim transcripts, are required of rate committee proceedings.

"5. Commission representatives may attend all rate bureau meetings; but that copies of correspondence and documents concerning all rate bureau meetings need not be filed with the Commission.

"6. A uniform system of accounts for rate bureaus will be promulgated.

"7. Rate bureaus are not prohibited from furnishing technical and professional services to other bureaus or nonmembers provided that the limitations expressed in the report are observed.

"8. Rate bureaus may not invest in another commercial business, whether related or unrelated to its primary function of processing and publishing rate and related matters for member carriers.

"9. Rate bureaus are prohibited from acquiring other rate bureaus without prior Commission approval.

"10. Rate bureaus may not be profitmaking enterprises.

"11. A carrier member of a bureau, which carrier is affiliated in any way with a shipper, may not serve on a bureau's board of directors, general rate committee, or any other committee which has an effect, either directly of indirectly, on the ratemaking function of the bureau without specific prior Commission approval.

"12. A maximum, period of 120 days is prescribed for the processing of proposals to final disposition.

"13. Public notice of proposals need not identify the proponent.

"14. Rate bureaus are prohibited from broadening the territorial or commodity scope of an individual rate proposal without prior adequate notice.

"15. Adoption of shortened special procedures involving proposals covering special docket applications are not warranted in this proceeding.

"16. Section 22 quotations require special bureau procedures limiting notification to members and the government agency.

"17. Docketing of rate bureau proceedings with respect to general rate increase proposals need not be mandatory.

"18. The Commission need not obtain and publish reports of the deliberations within the industry concerning the matter of general increases.

"19. The various rate bureaus need not join in seeking general rate increases.

"20. The various rail rate bureaus are not required to substantiate general increases on a regional basis only, but that regional costs should be presented in a more explicit manner.

"21. Rate bureaus are prohibited entirely from protesting proposals of carrier members, and are not merely limited to instances in which the proposed rate is less than long-term variable cost or any other specific instance.

"22. Rate bureaus are prohibited from discouraging independent action proposals of member carriers in any way, including the protesting of the filing of any rates pursuant to such action.

"23. Rate bureaus should be prohibited from discouraging members from publishing individual tariffs.

"24. Immunity from anti-trust laws shall be continued.

"25. Immunity from the anti-trust laws shall continue to be extended to agreements with respect to proposals of single-line movements.

"26. Additional legislation is not necessary, and need not be sought, to better effect the goals for which section 5a was enacted.

"We further conclude and find that all carrier respondents herein are required to file appropriate applications for approval of amendments to their respective agreements consistent with the requirements and standards set forth herein within 120 days."

Railroad Rate Committee Organizations

The Railroad Rate Committees and Sub-committees are:

Traffic Executive Association, Eastern Railroads
- General Freight Traffic Committee, Eastern Railroads
- General Freight Traffic Committee—T.O.F.C.
- Coal, Coke & Iron Ore Committee

National Diversion & Reconsignment Committee

National Perishable Freight Committee

Western Railroad Traffic Association Executive Committee

Western Trunk Line Committee
- Executive Committee
- Freight Traffic Managers Committee
- Colorado-Utah-Wyoming Committee
- Chicago Switching Committee
- Illinois-Indiana Coal & Coke Committee
- Illinois Rate Committee

Northern Lines Committee
Montana Lines Committee
Intermountain Committee

Southwestern Freight Bureau
Executive Committee
General Traffic Committee

Trans-Continental Freight Bureau
Executive Committee
Freight Traffic Managers Committee

Pacific Southcoast Freight Bureau Committee
Executive Committee
Freight Traffic Committee

North Pacific Coast Freight Bureau
Executive Committee
Freight Traffic Committee

Pacific Southcoast-North Pacific Coast Freight Bureau Committee
Executive Committee
Freight Traffic Committee

Southern Freight Association
Executive Committee
General Freight Committee
Coal & Coke Committee

Southern Ports Foreign Freight Committee

Uniform Classification Committee

While all rate commodities do not operate under the same rules of procedure, the general purpose of each rate committee is the same, and except for minor differences, such as, sending out information, issuance of rate advices, etc., all such committees function along similar lines.

All member roads are members of the General or Freight Traffic Managers Committees which meet periodically to consider such of the rate committee's reports as have met with disagreement. In addition, each Association has an Executive Committee made up of most and sometimes all of the chief traffic officers of the member lines which also meets at stated intervals.

Each major association also maintains what is known as a standing rate committee, auxiliary committee or research group. The function of this committee is to analyze the many rate proposals and other problems and recommend such changes as in their judgment would best serve the interests of the carriers and the interested shippers. Members of this committee, when necessary, often consult with the proponent and other shippers or investigate methods of manufacture at shippers' plants. The object is to ascertain all of the facts and after considering competing commodities, locations, markets, sources of raw material, and many other factors, to recommend approval or disapproval, or in certain instances a modified proposal for consideration by the carriers. Members of the standing rate committee do not vote. They merely recommend.

Members of standing rate committees have assigned to them a number of rate men who analyze proposals from the standpoint of present, proposed and competitive rates, and other information shown in the proposal to determine their correctness and also develop what are known as related files which disclose the historical background of the particular adjustment under consideration. This personnel also develops any additional data required by a member of the standing rate committee to whom the proposal has been assigned. Usually the proposals are divided among the members of the standing rate committee on the basis of the commodity involved, particularly as to those commodities that move in large volume, such as, grain, lumber, building material, etc. Otherwise the proposals are more or less equally divided to provide an equal share for each member.

Some rate committees also have commerce departments which consider the defense of new complaints. In most cases where the complaints involve but a few carriers, these cases are assigned to such carriers for defense with one usually assuming the chairmanship. Any necessary data may be prepared by the Association or by the member roads of interest, either in the offices of the Association or at member's offices. Where a case involves most of the member lines or if for other reasons the carriers so desire, the Association will furnish the witness.

Rate Committee Procedure

Shippers may file applications for changes in rates, rules, regulations, etc., through the carriers or the Association directly. These applications should contain complete information as to: (1) Commodity involved, (2) present classification, (3) origin point or points, (4) destination point or points, (5) present rate, rule, regulation or minimum weight, (6) proposed change desired, and (7) argument in support of application.

In the first place, it is the practice to issue a docket for each application for a rate, rule or regulation, etc., change that is received in the office of the committee. No attempt is made by the Committee to sift the merits of the proposal before docketing it, regardless of how obviously impracticable it may be. The docket thus made up gets wide publicity. In addition to the distribution to member lines, most of the committees maintain limited lists of shippers' organizations to whom the dockets are mailed, but the chief medium of dissemination is through the columns of Part 2 of the weekly *Traffic Bulletin,* published by the Traffic Service Corporation, Washington, D.C., in which the dockets of all of the rate committees are printed.

Following the publication of the docket, 7 to 14 days are allowed interested shippers in which to state their views, usually by letter, to the chairman of the committee. If a shipper or a group of shippers considers the material to be placed before the committee too extensive or too involved for presentation in the ordinary manner, application for hearing may be made, in which case the matter is set for hearing before the full committee, and a hearing bulletin is issued and published in the same manner as the docket bulletin.

Hearings before rate committees are informal affairs. The public is permitted to be present and appearances and testimony are accepted from anyone who may have information bearing on the subject under consideration.

In some instances the recommendations of the standing rate committee are distributed to the member lines for mail vote. Under this procedure, lines not responding within a limited period, usually from 5 to 10 days, depending on whether or not the proposal is of an emergency nature, are recorded as approving the recommendation of the standing rate committee. If objections are recorded within the voting period, the subject is listed for consideration at the next meeting of the general committee. In some cases, where in the judgment of the carriers or the committee, the subject is too important to clear by mail vote, it is listed for consideration by the general committee at one of its regular meetings. Sometimes, in the case of a real emergency, a subject may be considered at a special meeting of the general committee called for that purpose.

Whenever and however a final decision is arrived at, it is published by some of the committees in lists of dispositions. While this practice is not universal, the major number of the rate committees have adopted it, and these distribute the lists of dispositions in the same manner as their dockets. Thus it is possible for interested shippers to follow the course of a rate proposal, from the docketing to the disposition, by watching the columns of Part 2 of the weekly *Traffic Bulletin*. For example:

If the shipper is not satisfied with the action of the general committee,

TRAFFIC BULLETIN

PART 2

(Registered U. S. Patent Office)

January 26, 1980 ISSN: 0041-0659

PART 1 contains: Applications filed with the federal regulatory agencies, orders of such agencies, abstracts of tariffs filed with these agencies and embargoes issued by carriers.

EXPLANATORY: Advance notices of impending rate changes are contained in the dockets of the various rate committees. The time allowance for an appeal or request for hearing is as designated in the explanatory statement which appears following the name of the conference, association or committee; if a matter is set for hearing, notice of that hearing appears, giving the interested shipper opportunity to appear; notices of disposition of proposals are published regularly.

In addition to the rate committee dockets, the docket of the National Classification Board of the Motor Carrier Industry, is furnished as a special part of the Traffic Bulletin when issued.

The Traffic Bulletin is published weekly by The Traffic Service Corporation, 1435 G St., N.W., Suite 815, Washington, D.C. 20005.

RAIL

Dockets—Dispositions—Hearings

General Freight Traffic Committee—Eastern Railroads

The following submittals are docketed and if desired, shipper or carriers interested will be accorded hearing PROVIDED written request therefor is received by the Chairman within fourteen (14) days from the date of publication of this notice in the weekly *Traffic Bulletin*. If hearing is requested the date thereof will be subsequently announced. Action on any submittal will not be restricted to its exact scope, but may include other points of origin or destination or other commodities or recommendations varying from, but directly or indirectly related to the changes proposed. Rates, except as otherwise stated, includes X-368-A series increase and will apply in carloads, in cents per 100 pounds, except as otherwise stated. Where (C), (S) or (CS) is shown immediately following the submittal number, it means that it was initiated by (C) Carrier, (S) Shipper, (CS) Shipper, rate suggested by Carrier. The letters (ALT.) indicate that suggested rates will alternate with present rates.

Where the letter E is shown immediately following the submittal number, it denotes a proposition being handled under emergency procedure upon which *Public Hearings will not be held.*—WILLIAM J. BOLCH, Chairman, 1250 Broadway, New York, N.Y. 10001.

DOCKET BULLETIN

FOLIO NO. PD 80-04 - JANUARY 25, 1980

165S - 822 (C) Amend B&M Corp. F/T 8000-A ahd all other Agency and Individual Lines Switching Tariffs to provide a charge of $64.04 for turning cars to permit loading or unloading from a specific side. (Exhibit by Committee on request.)

201 - 686 (CS) Amend Items 301 and 303, TEA Tariff ICC 4043, Disability Rules to provide Disability due to shutdown of consignor's or consignee's facility.

F 305 - 132 (C) (SUPP. NO. 1 TO PD FOLIO NO. 79-42 DATED 10/27/79) PETROLEUM OR PETROLEUM PRODUCTS, Including Compounded Oils or Greases having a Petroleum Base as described in Items 77230 to 77340 of UFC, in packages, M/W 90M, not subject to Rule 24 of UFC or Exceptions thereto, to Newport News, VA from Baytown, TX 250¢ and Port Arthur, TX 236¢, not subject to X-357-A.

F 305S - 152 (CS) RESIDUE, Petroleum Resinous, By-Product of Ethylene Production, M/W R-35 and M/W provided therein for cars of not less than 20M gal. cap., BNLT 150M, Freeport, Chaison, TX and Plaquemine, LA to Cayuga, Ontario, Can., 265¢, including X-357-A, subject to X-311.

320S - 6607 (CS) Amend Item 4654-Series, WTL 3134-S to include Winchester, MA at the same rate and provisions as is applicable to Boston, MA on LITTER, Animal, consisting of Pelletized Vegetable Material, Etc., from Kansas City, KS.

320S - 6608 (CS) Amend Item 2870 of C&O Tariff 3000 by adding East St. Louis, IL as an additional destination on CALCIUM CHLORIDE, other than liquid at the same rates and provisions as presently apply to Granite City, IL from Ludington, MI.

F 320 - 6609 (C) CALCIUM CHLORIDE, other than liquid, in packages, Official Territory to SFA Territory. Revise CRC rates in Item 23110, TEA 2008-L as shown in statement which will be furnished by Committee on request.

320S - 6610 (CS) HYDROGEN, Liquefied, in T/C's, M/W-R35 except as to M/W which will be 40,000, Ceico, OH to Ora, SC, 600¢. (ALT.)

320E - 6611 (CS) CHLORIDE OF IRON, TC-R35 in 20,000 gal. car BNLT 190,000, Edge Moor, DE to Greensboro, NC 92¢, Lanett, AL 138¢.

F 320 - 6612 (C) ACRYLONITRILE, Chocolate Bayou and Texas City, TX to Dow, VA, 173¢ R-35 and M/W provided therein for cars of not less than 29M gal. cap., BNLT 190M, 201¢ R-35 and M/W provided therein for cars of not less than 20M gal. cap., BNLT 132M.

F 320 - 6613 (CS) MELAMINE, M/W 130M, not subject to R-24 of UFC or Exceptions thereto, to Detroit, MI from Donaldsonville, LA 228¢, from Avondale, LA 217¢.

F 321S - 1858 (C) Revise Commodity Rate Columns in various items of SWFB 2004-J, 2005-J, he, or a member road, may appeal to the executive committee with jurisdiction. If he is not satisfied with the action of that body and cannot persuade any road to take independent action, and if he still feels his present adjustment is unreasonable or in any other way in violation of any provisions of the Interstate Commerce Act, he may file a complaint with the Interstate Commerce Commission.

TRAFFIC BULLETIN

21510, 21515 & 21520, PSFB Trf. 3752-K.

1652 (C) PULPBOARD SHEETS, CL, Union Gap, Quadlock & Lacey, WA to Idaho Falls, ID. Prop. cancel Item 21590, PSFB Trf. 3752-K.

1653 (C) STARCH, CL, ID pts. to ID, OR, WA, BC pts. Prop. cancel Items 21790 to 21805, PSFB Trf. 3752-K.

1654 (C) CEMENT, CL, Lake Oswego, OR to Kennewick, WA. Prop. cancel Item 673, NPCFB Trf. 4735-A.

(X) 1655 (S) GRAIN, GRAIN PRODUCTS, CL, WA pts. to Arlington, Burlington, Fir and Stanwood, WA. Req. add Mitchell, WA at same rates as Stanwood, WA, Item 2225, NPCFB Trf. 4010.

(X) 1656 (S) LUMBER, Related Commodities, Transit at WA pts. Req. add Hoquiam, WA at same provisions as Aberdeen, WA, Item 1315, NPCFB Trf. 9834-J.

(X) 1657 (S) PHOSPHORUS, COFC, Don, ID to Tacoma, WA. Req. $1463.00 per container, NPCFB Trf. 7866-B.

\- - - - - - -

DISPOSITIONS

1541, 1565, Approved as docketed.

1422, 1477, Withdrawn.

North Pacific Coast Freight Bureau and Pacific Southcoast Freight Bureau

C. A. SPRENGELMEYER, Chairman
717 Market St., San Francisco, Calif. 94103

JOINT DOCKETS

(X) 8431 (C) LIQUID & ANTI-FREEZE CLEANING COMPOUNDS, TOFC, Salt Lake City, UT to Seattle, WA. Prop. 237¢, 80M in not more than 2 trailers, UP Trf. 7100.

8432 (S) LIME, Common, CL, from Caselton, NV. Req. to Coos Bay, North Bend, OR, 184¢, to Halsey, Springfield & Harrisburg, OR, 133¢, to Gardiner, OR, 156¢, to Toledo, OR, 154¢, 180M, PSFB Trf. 3701-X.

8433 (C) INK, Printing, CL, Betw. CA pts. & Portland, Seattle Grp. Prop. cancel 40M rates, Item 14800, PSFB Trf. 3701-X.

8434 (C) COPPER SCRAP, CL, ID, OR, WA pts. to McGill, NV. Prop. cancel Item 16700, PSFB Trf. 3701-X.

8435 (C) MAGNESIUM, Chloride of, CL, Portland Grp. to CA pts. Prop. cancel 40M and 60M rates, Item 19060, PSFB Trf. 3701-X.

8436 (C) PALLETS, Fibreboard or Paper, CL, Salinas, CA to ID, OR, WA pts. Prop. cancel Item 20920, PSFB Trf. 3701-X.

8437 (C) PAPER ROLL END PROTECTORS, CL, Betw. LA Grps. and West Linn, St. Helens, OR, Camas, WA. Prop. cancel Item 21020, PSFB Trf. 3701-X.

8438 (C) COCKS OR VALVES, Pipe, CL, Betw. CA pts. & ID, OR, WA pts. Prop. cancel 40M rates & Change 60M rates to "Between-And", Item 21740 and prop. cancel Item 21760, PSFB Trf. 3701-X.

8439 (C) STORAGE RACKS, CL, SF, Port Chicago Grps. to ID, OR, WA pts. Prop. cancel Item 22660, PSFB Trf. 3701-X.

8440 (C) RUBBER EXTENDER OR PROCESSING OIL, CL, CA pts. to Portland, Seattle Grps. Prop. cancel rates, except retain fr. SF, Port Chicago Grps., Item 23940, PSFB Trf. 3701-X.

8441 (C) SALT, Common, CL, Betw. LA Grps. & ID, OR, WA pts. Prop. cancel Item 24060, PSFB Trf. 3701-X.

8442 (C) SAND, CL, San Diego Grps. to Portland, Seattle Grps. Prop. cancel rates, Item 24260, PSFB Trf. 3701-X.

8443 (C) TABLES, Pool or Tennis, CL, Betw. CA pts. and ID, OR, WA pts. Prop. cancel Item 26860, PSFB Trf. 3701-X.

8444 (S) FOODSTUFFS, CL, Winnemucca, NV to Portland, OR. Req. 184¢, 60M, 168¢, 75M, 157¢, 100M, PSFB Trf. 3701-X.

8445 (Notice of Individual Action, Article III, Articles of Organization and Procedure, WRTA, following recd. from member line for immediate publication:) BUTANE OR ISOBUTANE GAS, CL, Bumstead, AZ to Cherry Point, WA. Amend Item 2762, PSFB Trf. 3727-J by extending expiration date to 3/1/80.

8446 (S) IRON OR STEEL SCRAP, CL, Thorne, NV to Seattle Grp. Req. 2695¢, PNT, 100M/80M, PSFB Trf. 3701-X.

8447 (S) BUTANE OR ISOBUTANE GAS, CL, Bumstead, AZ to Cherry Point, WA. Req. 240¢, R35/145M, Item 2762, PSFB Trf. 3727-J.

8448 (C) TRUNKS OR TRAVELING BAGS, CL, Betw. CA pts. & Portland, Seattle Grps. Prop. cancel Item 27320, PSFB Trf. 3701-X.

8449 (C) UREA SOLUTION, CL, CA pts. to Portland & Seattle Grps. Prop. cancel Item 27500, PSFB Trf. 3701-X.

8450 (C) CAMPER UNITS, FUEL TANKS, CL, Portland Grp. to SF, Port Chicago, LA Grps. Prop. cancel Item 27740, PSFB Trf. 3701-X.

8451 (C) AUTOMOBILES, Freight, CL, Velox, WA to SF, Port Chicago, LA Grps. Prop. cancel Item 27760, PSFB Trf. 3701-X.

8452 (C) WAX, Paraffine or Petroleum, CL, SF, Port Chicago Grps. to Woodwards Landing, BC. Prop. cancel Item 28360, PSFB Trf. 3701-X.

8453 (C) WOODENWARE, CL, CA pts. to Portland, Seattle, Vancouver, BC Grps. Prop. cancel Item 28640, PSFB Trf. 3701-X.

8454 (C) ZINC DUST, CL, LA Grp. to OR, WA, BC pts. Prop. cancel Item 29200, PSFB Trf. 3701-X.

8455 (C) DOLOMITE OR LIME ROCK, CL, Auburn, Newcastle, CA to Seattle & Vancouver, BC Grps. Prop. cancel rates, Item 33620, PSFB Trf. 3701-X.

8456 (C) BLACKS, CL, CA pts. to Bengal & Medicine Hat, AB. Prop. cancel Item 2040, PSFB Trf. 3739-V.

8457 (C) NICKEL ARTICLES, CL, Fort Saskatchewan, Thompson, AB to SF, Port Chicago, LA Grps. Prop. cancel Item 2540, PSFB Trf. 3739-V.

8458 (C) IRON OR STEEL ARTICLES, CL, to Metaline Falls, WA. Prop. from SF Grp. 187¢, from LA Grp., 219¢, 120M, Item 15360, PSFB Trf. 3701-X.

8459 (S) LADDERS, Wooden, CL, Betw. CA pts. & OR pts. Req. reinstate provisions, Item 43000, PSFB Trf. 3701-X.

8460 (S) OIL, VIZ.: CREOSOTE (Dead Oil of Coal Tar or Wood Tar). From Los Angeles Grps. Req. Seattle Grp., St. Helens, OR, 253¢, Bellingham, WA, 275¢, all R35/80M, Items 20140, 20160, PSFB Trf. 3701-X.

8461 (S) INSULATING MATERIALS, CL, from Fresno Grp. Req. Portland Grp. 572¢, Seattle Grp. 636¢, Central to Spokane Grps. 715¢, all in shipper owned cars not over 61'. Min. Wt. 24M, Item 32915, PSFB Trf. 3701-X.

8462 (C) COKE PETROLEUM CALCINED, CL, from Portland, OR to West Sacramento, CA. Prop. 1509¢, PNT, 190M, PSFB Tariff 3701-X.

\- - - - - - -

DISPOSITIONS

8376 Allow following Items (or Parts) to expire:

1) 1015, 36900, 36902, 37000, PSFB Trf. 3701-X.
2) 7170, 7340, 7350, 7395, PSFB Trf. 3702-Q.
3) 2360, PSFB Trf. 3739-V.
4) 3650, PSFB Trf. 3752-K.
5) 1110, 2867, PSFB Trf. 4491-A.
6) 15630, PSFB Trf. 7795-N.

(1)	(1)	(2)	(2)	(2)	(3)
8181	8343	8235	8323	8374	8183
8194	8346	8244	8329	8384	8337 &
8230	8370	8250	8332	8390	Sup. 1
8306	8378	8284	8336	8393	8380
8324	8389	8316 & Sups.	8339	8397	
8327	8404	1, 2 & 3	8341	8406	
8330	8411				
##8342	8412				

(4)	(4)	(4)	(4)	(4)	(4)
8251	8294	8338	8363	8382	*8398
#8268	8305	8344	8366	8383	8401
#8271	8321	8347	8367	8385	8405
8274	8322	8348	8368	8386	8408
#8275	8325	8349	8375	8394	8426
#8279	8331	8351	8381	8396	

(6)	(7)	(8)	(9)	(10)	(11)	(12)
8185, Corr.	8371	8179	8260	8288	8197	8373
		8372				

* In Part. # Balance.
(1) Approve as modified.
(2) Decline.
(3) Withdrawn & Cancelled.
(4) Approve as docketed.
(6) Decline & in lieu flag to expire in one year.
(7) Allow provisions to expire.
(8) Extend provisions 1 year.
(9) Decline Orig.: Approve Sup. 1.
(10) Decline Orig.: Approve Sup. 1 as modified.
(11) Decline Orig. & Sup. 1; Approve Sup. 2 as modified.
(12) Extend provisions 6 Months.
\## Supercedes publicizing in 1-5-80 Traffic Bulletin Issue.

\- - - - - - -

NOTICE

BROADENED COMMODITY OR TERRITORIAL SCOPE

8333 FIREBRICK, PAVING BRICK, FIRE BRICK SHAPES, CL, from Abbotsford, BC. Prop. to SF Grp. 249¢, to LA Grp. 287¢, 60M, PSFB Trf. 3701-X.

Approve on Brick, fire, clay or other than clay from Vancouver, BC Grp. to SF, Port Chicago Grps. 280¢, to LA Grp. 345¢ to San Diego Grp. 365¢, 60M, PSFB Trf. 3701-X, rates not applicable via CPW.

8335 LIGNIN LIQUOR, CL, Lebanon, OR to Ger-

Motor Carrier Rate Committee Organizations

Rate committees play a very important role in the motor carrier industry also. The rate committee system provides a procedural framework for the orderly handling of proposals for changes in tariff provisions, guaranteeing adequate notice to the public and to competing carriers.

There are numerous motor carrier rate associations throughout the country, and all of the major bureaus are quite uniform insofar as rate committee procedures are concerned. The principal rate associations include:

Central and Southern Motor Freight Tariff Association

Central States Motor Freight Bureau

Eastern Central Motor Carriers Association

Interstate Freight Carriers Conference

Middle Atlantic Conference

Middlewest Motor Freight Bureau

New England Motor Rate Bureau

Niagara Frontier Tariff Bureau

Pacific Inland Tariff Bureau

Rocky Mountain Motor Tariff Bureau

Southern Motor Carriers Rate Conference

The organizational structure of the committees is quite similar for all of the major motor carrier associations. A Standing Rate Committee, usually made up of a chairman and two members, initially considers the proposals. This committee also has a secretary, or one of the members may serve in the dual role of secretary and member. All persons connected with the Standing Rate Committee are full time employees of the association. Superior to the Standing Rate Committee is the General Rate Committee, consisting of executives of the carriers desiring representation on this Committee. The primary function of the General Rate Committee is to act as an appellate body with respect to the decisions of the Standing Rate Committee. Some associations do not use the term General Rate Committee, having assigned a different title—for example, the Central States Motor Freight Bureau refers to its appeal committee as the Central Committee.

Procedure for Tariff Changes

Proposals for changes in tariff provisions may be submitted by shippers, carriers or by the association staff. When the proposal is received at the association, it is docketed for consideration by the Standing Rate Committee and publicized in Part 2 of the weekly *Traffic Bulletin,* Transport Topics, or through the association's own docket service, usually ten days or more before the hearing date. Excerpts from Part 2 of the Traffic Bulletin on page 251 indicate the manner in whch motor carrier proposals are publicized. Prior to the hearing, each proposal scheduled for consideration is analyzed and evaluated by the Standing Rate Committee. The information submitted by the proponent is checked for completeness and accuracy and additional data are sought from the proponent, if necessary. The analysis of the proposal includes such factors as the rates of competitive carriers, market competition, carrier or carriers presently handling the shipments, and the probable impact of the proposed adjustment on rates applying on the same commodity between other points, as well as the effect on the rate levels maintained on commodities with similar transportation characteristics in the territory. Any interested person who, prior to the hearing date, requests to be heard, may appear at the hearing in support of, or in opposition to, the proposal. After the hearing on the proposal the Standing Rate Committee disposes of it by majority vote. The power of the Standing Rate Committee is usually limited to making recommendations—it may recommend approval, disapproval, approval on a modified basis, or it may refer the matter directly to the General Rate Committee. If additional time is needed to acquire important information or for other good reasons, action on the proposal is deferred to the next meeting of the Committee. The Standing Rate Committee's disposition of the proposal is publicized in the same way as was the original proposal. In addition, the proponent and persons who have expressed an interest in the proposal are usually notified by mail.

While the Standing Rate Committee is limited to making recommendations, if no objections to its recommended disposition are received within a specified time period, the recommended disposition is accepted by the General Rate Committee without further consideration. If objections are received within the prescribed time limit, the proposal goes to the General Rate Committee for hearing. This hearing, too, is publicized and interested persons may appear to present their views. After all parties have been heard, the Committee discusses the proposal and reaches a conclusion on the basis of a majority vote. The action of the General Rate Committee is final, although the proponent of a rejected proposal may refile it or request reconsideration. In most cases, this would be futile unless new developments had occurred in the meantime which might change the Committee's thinking. Action by the General Rate Committee is the final step in rate association procedure, but a dissatisfied proponent always has the right to go beyond the association procedure and file a complaint with the Interstate Commerce Commission attacking the existing rate, rule or other traffic provision.

In case of carrier proposals, it is not necessary that the entire rate committee procedure be followed, or, in fact, that it be followed at all. This is because each member carrier has the free and unrestrained right of independent action. In the exercise of this right, all the carrier need do is instruct the association to establish a specified rate, rule or regulation for its account. An announcement of this independent action is then circulated among the other member carriers in order to permit them, for competitive reasons, to also have the rate, rule or regulation published for their account. Independent action can be taken at any time before, during or after the regular rate committee procedure. It most commonly occurs in those territories in which there are large numbers of motor carriers and competition for traffic is extremely keen.

CHAPTER 16

TECHNICAL TARIFF INTERPRETATION

General

In discussing the subject matter of this chapter, we are brought into contact with some of the peculiarities of traffic work. To the uninitiated it may appear as though the entire field of traffic is technical and that a study of the particular subject of technical tariff and rate interpretation would be considered in the light of an attempt to "gild the lily."

The word "technical," however, is often used, or rather misused, by less experienced tariff and rate people whenever a problem presents difficulties of interpretation and solution. In other words, by far the majority of involved traffic situations are not technical in the true sense of the word, and the hue of inconsistency which gives a different, and very often erroneous, color to a given situation is produced wholly by a conflict of inexperienced reasoning and a lack of proper analysis on the part of the persons exercising the judgment. In such instances, the properly trained person is able to immediately differentiate between genuine technicalities in tariff and rate work and mere difficulty in interpretation, and finds that only the application of keen traffic sense and a more mature judgment in the legal phases of the profession are necessary in order to reach the proper conclusion.

In the interpretation of tariffs it must be remembered that tariff provisions have the force and effect of statutes, and must be interpreted according to the plain terms of the provision; and while the Commission has followed the principle that if tariffs are doubtful in meaning, the construction will be applied that is most favorable to the shipper, they have also followed the principle that the fair and reasonable construction will outweigh a strained or unnatural construction of the language. (*Blish Milling Co. v. Alton R. Co.,* 232 I.C.C. 331.)

During the years since the passage of the Act to Regulate Commerce, hundreds and hundreds of situations involving the interpretation of tariffs have been submitted to the I.C.C. and the courts and it would seem that every possible question must surely have been resolved, and yet, questions continue to be raised. Descriptions, rules and regulations, words and phrases written by carriers are interpreted or misinterpreted by shippers as something other than what was intended by the carrier and the result, litigation. Nevertheless, over these many years a host of tariff interpretation principles have evolved and it is these accepted principles which will be discussed in this chapter. Based on decisions of the Commission and the courts, they constitute a firm foundation on which new situations may be analyzed and new decisions structured.

Tariff Rate the Only Legal Rate

The Interstate Commerce Act, after specifying in Sections 6(1) (railroads), 217(a) (motor carriers), 306(a) (water carriers) and 405(a) (freight forwarders), that carriers must file their tariffs with the Commission and keep them open to public inspection, provided in Section 6(7), 217(b), 306(c) and 405(c), for positive adherence to these tariffs in the following language:

> "Nor shall any carrier charge or demand or collect or receive a greater or less or different compensation for such transportation of passengers or property, or for any service in connection therewith, between the points named in such tariffs than the rates, fares, and charges which are specified in the tariff filed and in effect at the time; nor shall any carrier refund or remit in any manner or by any device any portion of the rates, fares, and charges so specified, nor extend to any shipper or person any privileges or facilities in the transportation of passengers or property, except such as are specified in such tariffs."

In interpreting this provision of the law, the Commission, in a very early case, *A.J. Poor Grain Co. v. C.B. & Q. Ry. Co., et al,* 12 I.C.C. 418, 422, stated:

> "This clause (Section 6), heretofore ineffective in sustaining the published rate and preventing rebates, is now supported by strong provisions elsewhere in the same act and by provisions even more drastic in the so-called Elkins act as amended, which define any departure or variation from the published schedule as a misdemeanor punishable by severe fines and, under some circumstances, by imprisonment. Under the force of these enactments the published rate has taken on a fixed and rigid character that does not permit it to yield either to importunity or to mistake. And the giving of rebates, if not absolutely at an end in isolated cases, is at least no longer a practice in interstate transportation. In this respect, the published rate has become a protection to shippers and to carriers alike. Regardless of the rate quoted or inserted in the bill of lading, the published rate must be paid by the shipper and actually collected by the carrier. *The failure on the part of the shipper to pay or the carrier to collect the full freight charges, based upon the lawfully published rate for the particular movement between two given points constitutes a breach of the law and will subject either one or the other, and sometimes both, to its penalties.* In other

words, although a rate between two given points is established in the first instance by the voluntary act of the carrier in filing and posting it in the manner required by law, and in the same manner may be cancelled by the carrier and another rate substituted, nevertheless, *when once lawfully published, a rate, so long as it remains uncancelled, is fixed and unalterable either by the shipper or by the carrier as if that particular rate had been established by a special act of Congress. When regularly published, it is no longer the rate imposed by the carrier, but the rate imposed by the law.* The law governing transportation between the two points for which that rate was established requires that rate to be paid by the shipper and collected by the carrier, and any departure from it, either by the carrier or the shipper, is as much a crime as it would be if the rate had been fixed by specific enactment. Not even a court may interfere with a published rate or authorize a departure from it, when it has been established voluntarily by the carrier. *Texas & Pacific Railway Co. v. Abliene Cotton Oil Co.*, 204 U.S. 426 ***

"A rate may be lawful in the sense that it is the regularly published rate and therefore the only rate under which traffic may lawfully move, and yet at the same time be unlawful in the sense that it is excessive and unreasonable in amount. Its lawfulness as the published rate is to be tested by the mere inspection of the schedules on file with the Commission; and if found to have been published in conformity with the requirements of law, *that rate must in all cases be charged and actually collected by the carrier even though it may be excessive.* Whether or not it is lawful in the sense of being excessive depends upon all the circumstances and conditions that are recognized as having a legitimate influence in rate making. But in no event may a carrier agree with a shipper that the published rate, under which shipments have been made, is too high and undertake to adjust it by collecting a less rate. The publication fixes and makes the rate inflexible and unalterable, as heretofore explained, and that rate must be applied until changed in the manner required by law. Nor may a carrier, when a rate has been duly cancelled and a lower rate made effective, voluntarily refund from the freight charges already collected, and thus give the benefit of the new rate on shipments that moved under the previous higher rate. This Commission alone has jurisdiction to authorize a refund to be made from the charges collected on previous shipments. And it can give such authority only upon an affirmative finding that such rate was excessive and unreasonable, and therefore unlawful."

In *R.H. Coomes, et al, v. C.M. & S.P. Ry. Co., et al,* 13 I.C.C. 192, 193, the Commission reiterated this interpretation in the following concise and emphatic language:

"All provisions of the act must be read and construed in the light of each other, so as to give due effect to the whole. No proposition respecting the requirements of this Act is more clearly and firmly settled than that rates duly established as therein required are absolutely binding upon carriers and shippers alike until lawfully changed as also therein provided, but this does not render nugatory the provisions of the first section that all rates must be reasonable and just, nor the subsequent provisions authorizing the Commission upon proper showing to award damages resulting from violations of the Act. It follows that, although a rate is by the terms of the law binding upon all so long as it remains in effect, such rate may, nevertheless, upon proper procedure, be found and declared to be unlawful in that it is unreasonably high or unduly discriminatory, and become, in respect to shipments made while the unjust rate was in effect, the basis of an award in damages. To hold otherwise would be to make the mere establishment of rates by a carrier conclusive of their reasonableness and justness while in effect. Poor indeed would be the plight of shippers who have been compelled to pay excessive rates under such interpretation of the law. While the establishment of rates by the carrier in the manner required by law fixes the standard or lawful rates for the time being and so long as such established rates are in effect, this standard is by no means conclusive of their reasonableness and justness."

Also, in the case of *Hocking Valley Railway Company v. Lackawanna Coal & Lumber Company,* 224 Federal Reporter, 930, 932, the United States Circuit Court held as follows:

"While that rate stands (tariff rate) in the published tariffs of the carriers, it is the only legal rate, and binds shipper and carrier alike. The regulating statute on this point is explicit, and neither party can avoid its provisions. This has been repeatedly held by the Supreme Court in a long line of cases from *Gulf, Colorado & Santa Fe v. Hefley,* 158 U.S. 98, 15, Sup. Ct. 802, 39 L. Ed. 910, to *L. & N. Railroad Company v. Maxwell,* 237, U.S. 94, 35 Sup. Ct. 494, decided April 5, 1915."

Section 10761(a) of the present statute, P.L. 95-473, reads as follows:

"(a) Except as provided in this subtitle, a carrier providing transportation or service subject to the jurisdiction of the Interstate Commerce Commission under chapter 105 of this title shall provide that transportation or service only if the rate for the transportation or service is contained in a tariff that is in effect under this subchapter. That carrier may not charge or receive a different compensation for that transportation or service than the rate specified in the tariff whether by returning a part of that rate to a person, giving a person a privilege, allowing the use of a facility that affects the value of that transportation or service, or another device."

The language is slightly different from that of the Interstate Commerce Act but the substance is the same, namely that the tariff, as long as it is in effect, is the final word in the application of freight charges and all rates, rules and regulations which it contains must be strictly adhered to. This is the case even to the extent that the rates or other provisions of the tariff may be unreasonable, discriminatory, or otherwise in violation of the law. If the Commission, upon proper presentation and procedure, subsequently orders a change in the tariff or an award of reparation because of such unlawful feature, that is another matter; but, while it is in force, deviation from the tariff as printed will not be permitted under any pretext. This rule is undeniably strict, and it obviously may work hardship in some cases, but it embodies the policy which has been adopted by Congress in the regulation of interstate commerce in order to prevent unjust discrimination.

It is in the application of this principle of strict adherence to the tariff rate that we encounter the seeming paradox "legal but unlawful." A rate that is duly published and filed with the I.C.C. is the "legal" rate and must be applied. It may, however, violate some other provision of the statute and therefore be unlawful at the same time. In *U.S. v. N.P. Ry.*, 301 I.C.C. 581, the Commission held:

> "Under the Act, the rate duly filed with the Commission is the legal charge that may be made. Even the fact that a rate named in a published tariff is unreasonable and has been so declared affords no justification for a shipper paying or a carrier receiving other than the tariff rate."

Tariffs Must Be Construed Strictly According to Their Language

It is also well settled that once a tariff is filed with the Commission, it must stand as it is written. What it actually says, not what the framers of the tariff intended to say, governs its application.

While there are many decisions to support this statement, the following expression of the Commission, as contained in *Newton Gum Company v. C.B. & Q. Railroad Company, et al,* 16 I.C.C. 341, 346, will be found illustrative of the reasons underlying this feature.

> "The law compels carriers to publish and post their schedules of charges upon the theory that they will be informative. The shipper who consults them has a right to rely upon their obvious meaning. He cannot be charged with knowledge of the intention of the framers or the carriers canons of construction or of some other tariff not even referred to in the one carrying the rate. The public posting of tariffs will be largely useless if the carriers interpretation is to be dependent upon tradition and the arbitrary practices of a general freight office. *This Commission has long since repudiated the suggestion that railroad officials may be looked to as authority for the construction of their tariffs* [*see Hurlburt v. L.S. & M.S. Railway Company,* 2 I.C.C. 81]. We quote from Judge Cooley's opinion in that case:
>
> " 'A classification sheet is put before the public for its informa-

tion. It is supposed to be expressed in plain terms, so that the ordinary business man can understand it, and, in connection with the rate sheets, can determine for himself what he can be lawfully charged for transportation. The Committee who prepared this classification have no more authority in construction than anybody else, and they must leave the document, after they have given it to the public, to speak for itself.' "

In dealing further with this subject, the Commission in *American Cotton Waste & Linter Exchange v. B. & O. R.R. Co., et al,* 169 I.C.C. 710, 712, said:

"Defendants contend that the item covering mattresses was intended for new mattresses and does not, therefore, embrace old or used mattresses. That contention is unsound. Tariffs are construed according to their language and the intention of the framers is not controlling. The language used in that item neither restricts it to new mattresses nor excludes old mattresses therefrom."

The Commission has also indicated that tariffs should be clear, simple, and definite as is evidenced by the following quotations, contained in the case of *H.B. Pitts & Sons v. St.L. & S.F. Railroad Company, et al,* 10 I.C.C. 684, and *N.O. Nelson Manufacturing Co. v. Mo. Pac. R.R. Co., et al,* 153 I.C.C. 272, 275, respectively.

In the first case cited, the Commission stated:

"It is the duty of railroad companies under the Act to Regulate Commerce, to print, publish and file tariffs showing rates which are so simplified that persons of ordinary comprehension can understand them."

And in the second case, the Commission held as follows:

"The law not only requires that tariff provisions shall be reasonable, but also that they shall be plainly stated."

The Commission has from time to time condemned certain tariff provisions brought to its attention because they were lacking in this respect, as evidenced by the following.

In the case of *Mixed Car Dealers Association, v. D.L. & W. Railroad Company, et al,* 33 I.C.C. 133, 144, the Commission said:

"The statement that a rate or charge shall not be less than a certain amount is in no sense a clear and definite statement of what the charge will be."

And in *Tri-State Pipe Co. v. Abilene & Southern Ry. Co., et al,* 234 I.C.C. 55, 56, the Commission held:

"That part of the restrictive clause in the tariff, 'returned to manufacturers,' is indefinite. According to defendants' witness, it was intended to apply to pipe manufacturers which, at that time, were the only shippers of thread protecting rings. The intention of the maker of a tariff, of course, is not controlling*** "

Therefore, as the law specifically provides that the tariff is the final word in the assessment of freight charges and as the Commission has indicated that it is the duty of the carriers to publish and file their tariffs in such manner that persons of ordinary comprehension can understand them, the inexperienced individual, proceeding from these premises, might very naturally raise the question as to why there should ever be any technicalities in tariff interpretation. The argument might very well be advanced that if the user understood the tariffs, was acquainted with the procedure and rules governing their application, and read with a clear and open mind, but one solution could be arrived at—the correct one. However, such a presumption would be predicated upon a hypothesis as elusive as the most vagrant will-of-the-wisp—a hypothesis based on the assumption that tariffs are in every instance perfect and that the persons reading and interpreting them are also perfect in their conception and understanding of the information they contain. If that were a fact, then there would be no technicalities and no misinterpretations, but as long as the minds of those who compile and interpret the multitude of publications which are issued yearly run in different grooves and are guided by different thoughts and motives, the finer shades of meaning which are of necessity injected in some tariffs, or errors and omissions, will very often resolve themselves into real technicalities.

A case in point is *Swift & Company v. Alton R. Co.,* 263 I.C.C. 245; 258 I.C.C. 103. Swift & Co. shipped numerous carloads of fresh meats and mixed carloads of fresh meats and packing-house products, from Missouri River points to Cincinnati, Ohio. Charges were collected on basis of aggregates of specific proportional commodity rates to and from east-bank Mississippi River crossings. The use of the aggregates of these specific proportional commodity rate factors on this traffic was authorized under specified conditions by the aggregate-of-intermediates rule contained in the tariffs which published joint single-factor class rates from the origins to Cincinnati.

Swift & Co. argued that Cincinnati was intermediate to Louisville, Ky., over the routes via which the shipments moved to Cincinnati, and that, consequently, lower joint commodity rates to Louisville were applicable on the shipments to Cinncinati under the intermediate destination point rule contained in the tariffs naming such joint commodity rates to Louisville.

The destination intermediate-point rule in the tariffs publishing the joint commodity rates to Louisville (Agent Kipp's I.C.C. A-2109 and A-3112) read as follows:

> "Subject to the provisions of Notes *** 3 and 4 below, to any point of destination to which a commodity rate on a given article from a given point of origin and via a given route is not named in the tariff, which point is intermediate to a point to which a commodity rate on said article is published in this tariff via a route through the intermediate point over which such commodity rate applies from the same point of origin, apply to such intermediate point from such point of origin and via such route the commodity rate in this tariff on said article to the next point beyond to which a commodity rate is published herein on that article from the same point of origin via the same route.
>
> "**Note 3**—If the class rate on the same article via the same route to the intermediate point produces a lower charge than would result from applying the commodity rate under this rule, such commodity rate will not apply.
>
> "**Note 4**—If there is any other tariff a commodity rate on the same article to the intermediate destination point applicable over the same route from the same point of origin, the provisions of this rule are not applicable to such intermediate destination point."

The following aggregate-of-intermediates rule appeared in the tariffs which named the joint class rates to Cincinnati: (Agent Jones' I.C.C. Nos. 2920 and 3243).

> "If the aggregate of separately established (joint, local and/or proportional) rates applicable on interstate traffic contained in tariffs *** applicable via any route over which the through rates published in this tariff *** apply, produces a lower charge on a shipment than the rates published in this tariff, *** such aggregate of rates will apply via all routes over which the rates shown in this tariff *** are applicable, and the through rate published in this tariff *** has no application to that shipment."

In its findings in this case the Commission stated that Notes 3 and 4 of the intermediate rule provided definite limitations on the application of that rule. As the joint class rates maintained by defendants from origins to Cincinnati were higher than the joint commodity rates to Louisville, Note 3 does not preclude application of the intermediate rule on this traffic. Note 4 limits the application of the intermediate rule so that the provisions of that rule are not applicable to any intermediate point "if there is in any other tariff a commodity rate" on the same article to such intermediate point. The rule itself is limited so that it only applies "to any point of destination to which a commodity rate *** is not named in this tariff." Thus, the intermediate rule as qualified by Note 4 clearly has no application to any intermediate point in situations where a commodity rate is published to such intermediate point either in the tariff containing the intermediate rule or in any other tariff.

The application of the intermediate rule turns squarely on the answer to this question: Do the aggregates of specific proportional commodity rates to Cincinnati, constitute commodity rates "in any other tariff" within the meaning of Note 4 of the intermediate rule? Complainant argued that these aggregates of proportional commodity rates to Cincinnati were not joint rates, and that since the individual proportional commodity rate factors appeared in separate tariffs, the resulting aggregates did not constitute commodity rates "in any other tariff" within the meaning of Note 4. That interpretation of the tariffs, said the Commission, ignores certain basic principles of tariff construction.

In its discussion of these principles, the Commission states, that in determining exactly what kinds of rates were published to Cincinnati, for the purpose of construing Note 4 of the intermediate rule, it is necessary to have in mind certain well-settled rules of tariff construction under which, in the absence of tariff provisions to the contrary, a joint commodity rate applies under section 6 (7) of the Act to the exclusion of all other kinds of rates; similarly, a joint class rate applies to the exclusion of combination rates. Under these rules, the joint class rates to Cincinnati would have applied to the exclusion of the proportional commodity rate factors to that point except for the tariff authority contained in the aggregate rule, previously quoted herein, published in the tariffs naming the joint class rate to Cincinnati. This means that the proportional commodity rate factors *as such* had no application to the shipments. But since the aggregates of such proportional commodity rate factors produced lower charges than were produced by the joint class rates, the aggregate rule provided that:

> "Such aggregate of rates will apply via all routes over which the rates shown in this tariff *** are applicable, and the through rate published in this tariff *** has no application to that shipment."

Referring to the tariff authority just quoted, the Commission argued, it should be emphasized, first, that it is the *aggregate* of the separately established rates that applies as the through rate, and, second, that such *aggregate* applies over all routes over which the joint class rates apply. The aggregate rule does not authorize the alternative use of the proportional commodity rate factors in such manner as to justify the conclusion that those proportional rate factors were applicable to Cincinnati by virtue of tariff authority contained in the tariff naming the individual factors. Instead, the tariff authority for the use of "such aggregate of rates" is contained in the aggregate rule as evidenced by the fact that such aggregate of rates applies over all routes over which the joint class rates apply irrespective of whether or not the proportional commodity rate factors as such apply over such routes.

The following definition of the term "joint rate" appears at page 1 of *Tariff Circular 20:*

> "The term 'joint rate,' as used herein, is construed to mean a rate that extends over the lines of two or more carriers and that is made by arrangement or agreement between such carriers and evidenced by a concurrence or power of attorney."

The tariff authority for the application of the aggregates of proportional commodity rates to Cincinnati in lieu of the joint class rates to that point, that is the aggregate rule, was published in a single tariff which, although predominantly a class tariff, also contained commodity rates, and it (the rule) was concurred in by carriers whose lines are embraced in the through routes to Cincinnati. Consequently, the through rates to Cincinnati made by the aggregate rule are rates (in the language of the tariff circular) "that extend over the lines of two or more carriers" and such rates are "made by arrangement or agreement between such carriers and evidenced by concurrence or power of attorney."

Also, Tariff Circular 20 recognizes only two categories of rates, namely, class rates and commodity rates. These proportional commodity rate factors to and beyond the river crossings appeared in tariffs which published only commodity rates on the particular commodities referred to in those tariffs, including the commodities here considered. These proportional commodity rate factors added together for the purpose of obtaining the through rate under authority of the aggregate rule, of course, did not lose their character as commodity rate factors; and the resulting through rate was a commodity rate for the obvious reason that it was an aggregate of commodity rate factors. This conclusion is consistent with the findings of division 4 in *Morris Buick Co., Inc. v. Grand Trunk W. R. Co.,* 299 I.C.C. 109.

In conclusion, the Commission said, that a finding in this proceeding that the aggregates of specific proportional commodity rates to Cincinnati constituted *commodity rates in another tariff* within the meaning of Note 4 would not be inconsistent with the decision in *Rogers v. Cincinnati, N.O. & T.P. Ry. Co.,* 258, I.C.C. 553. In that proceeding the Commission found that combination rates composed of local commodity rates to Cincinnati and a proportional commodity rate beyond were not "commodity rates" within the meaning of Note 4. But the commodity rate factors in that proceeding were applicable on the through traffic by virtue of tariff authority contained in the tariffs publishing the individual factors. Such combinations did not result in joint rates in any sense of the word. There was no aggregate rule involved in that proceeding. Therefore, neither the rates themselves nor the tariff authority for the use of the aggregates appeared in a single tariff, as contrasted with the situation in the instant proceeding as previously detailed herein.

Inasmuch as these aggregates of commodity rates to Cincinnati are composed of specific proportional commodity rate factors, and since the tariff authority for applying such aggregates on this traffic appeared in a single tariff, it is of no particular significance that the mechanics of arriving at the aggregate or single sum, under the tariff situation disclosed in the Swift case, required the addition of two amounts appearing in different tariffs—no more so than in the many comparable situations where a determination of charges based on orthodox single-factor joint rates requires that tariffs other than the one containing the rate be consulted, such as master tariffs providing for general increases or reductions, the governing classification, transit tariffs, et cetera. The Commission held that the controlling facts here are:

(1) That the aggregates of the rates to and from east-bank Mississippi River crossings are composed of specific proportional commodity rate factors which aggregates have been recognized in other proceedings as constituting the applicable rates to Cincinnati;

(2) That the tariff authority for applying such aggregates to Cincinnati appears in a single tariff;

(3) That such tariff authority was concurred in by carriers whose lines are embraced in the through routes to Cincinnati; and

(4) That it was only by virtue of such tariff authority appearing in a single tariff that the aggregates of such factors could be applied on this traffic to Cincinnati.

There are times when a decision handed down by the Commission or a court seems completely contrary to all of the principles and precedents which have evolved and been accepted for many years. This Swift case is one of those decisions. The dissenting opinion written by Commissioner Aitchison and concurred in by Commissioners Porter and Lee, seems more in keeping with sound principles of tariff interpretation. In view of the thoroughness and lucid manner in which the complicated subjects of intermediate application and the aggregate of intermediates provision are discussed in this dissenting opinion and the contribution it makes to an understanding of tariff interpretation, the entire text is being quoted.

"Aitchison, Commissioner, dissenting:

"In reaching the conclusion that the intermediate rule has no application in this situation, the report relies on a novel theory of tariff interpretation, which in my view is unsound and will retard rather than encourage tariff clarity and simplification.

"The intermediate rule involved provided that the commodity rate published in the tariff to the next more distant point applied to the intermediate point, unless there was in some other tariff a commodity rate on the same article from the same origin to the intermediate point. The effect of the report is to hold that there was such a commodity rate in another tariff, and that therefore the commodity rate to the next more distant point did not apply to the intermediate point. In the process the report departs from time-honored and salutary rules of tariff construction, and by devious means arrives at the conclusion that two separate proportional rates, published in two separate tariffs, are "a commodity rate in any other tariff," within the meaning of note 4 of the intermediate rule.

'Assuming for the sake of argument that the provisions of note 4 are ambiguous, the report resolves the ambiguity in favor of the maker of the tariff and against the shipper, contrary to sound principles followed in a long line of decisions both by the courts and this Commission. When two plausible constructions of a document are possible, one lawful and the other unlawful, it is a sound presumption of law that the former intended that construction which produces lawful results. And yet under the theory of the report, the higher combination to the intermediate destination would apply in preference to the lower joint rate to the farther distant point, in violation of the long and-short-haul clause of the fourth section of the act, the very result the intermediate rule is intended to prevent.

"The tardiness of shippers, railroads, and this Commission in recognizing the rate changes which resulted from the insertion of the interediate rules in the joint tariff has no proper bearing on the interpretation of the tariff. As early as 1935, at least one railroad, with a route through Cincinnati which might have been used on traffic from West to Louisville, restricted the Louisville rate so as not to apply over its route through Cincinnati. Other carriers thereafter by tariff publications similarly restricted the application of the Louisville rate. But, despite this clear notice of the effect of the intermediate rule, that rule was in the tariff for over 10 years before adequate action was taken to close the routes by way of Cincinnati. I cannot join in giving encouragement to the present all too prevalent practice of the railroads which insert the intermediate rule in their tariffs to obtain immunity from prosecution for violation of section 4 of the act with no proper regard for the effect on the rates, and then resort to strained construction of the tariffs to avoid the results of their heedlessness.

"This decision sets up another basis for constructing through rates, which detracts from the finality and definiteness of the joint through commodity rate and injects a new cause of conflict and confusion into an already complex tariff situation. Years ago many carriers provided in their tariffs 'a specific basis for constructing through rates' which caused so much difficulty and conflict that our tariff rules were finally revised so as to require the cancellation of such arrangements. In resurrecting this cause of past confusion, the report is a step backwards.

"An attempt is made in the report to confine the application of the higher combinations in lieu of the lower joint rates, to instances in which those combinations are composed of proportional factors, but I doubt if the theory on which the report is based can be so confined.

"The report first makes three preliminary findings:

"(1) An aggregate displacing a joint class rate is a joint rate;

"(2) if composed of two commodity factors it is a commodity rate; and

"(3) notwithstanding that the factors are physically published in separate tariffs, such aggregate must be considered to be in class tariff (and therefore in a single tariff) merely because the formula which calls for the use of separately published factors is perforce in the class tariff.

"If these three conclusions are sound, obviously it matters not whether the factors are both proportional rates, both flat rates, or one flat rate and one proportional rate, because note 4 has no more reference to proportional rates than to flat rates. Note 4 deals only with a commodity rate in another tariff. If separately published commodity rates in different tariffs can be considered to be in one tariff merely because the formula which produces the aggregate is in one tariff (a conclusion which I consider to be unwarranted), it follows that such aggregate will displace the commodity rate to the more distant point whether it be composed of proportionals, flat rates, or one or more of each kind of rate.

"The aggregate rule clearly states that two or more separately published rates displace the joint overhead rates if lower charges result. It reads in part: *If the aggregate of separately established joint, local and/or proportional rates *** such aggregate of rates will apply ****. To say that an aggregate of rates is one rate completely disregards the plain wording of the rule. The report uses the word 'aggregate' as a synonym for a 'commodity rate.'

"The words 'and the through rate published in this tariff *** has no application to that shipment' as used in the aggregate rule serve a definite purpose which has been entirely overlooked in the report. The effect of those words is to eliminate conflicting rates—in this instance, to set aside the joint class rate and permit the application of the aggregate of rates. If an aggregate of separately established proportional commodity rates made applicable by virtue of an aggregate rule were, in fact, a commodity rate, there would be no need for the quoted provision in the aggregate rule, because a commodity rate would automatically remove the application of the joint class rate.

"In raising the aggregate of the separate proportionals to the dignity of a joint commodity rate within the meaning of note 4, the report brings about two important results. The joint commodity rate would not only displace a specific rate under the intermediate rule at the intermediate point, but it would also create a conflict in rates where a local or joint commodity rate is also specifically published from the origin to the farther distant destination. For example, the same combination of proportionals as charged to Cincinnati was also published from these origins to Louisville. By reason of the aggregate-of-intermediates rule contained in the class tariff publishing the class rate to the more distant point, Louisville, a conflict in commodity rates would be created to the more distant point as well as to Cincinnati. In such instances the lowest charge resulting from the application of either of the rates would be applicable. See *Chicago, I & L. Ry. Co. v. International Milling Co.*, 33 Fed. (2d) 636, 43 Fed. (2d) 93. A construction of the aggregate and intermediate rules which would result in conflicts and unlawfulness in the establishment of rates should be avoided.

"In emphasizing the decision of division 4 in *Morris Buick Co., Inc., v. Grand Trunk W. R. Co.*, 229 I.C.C. 109, the report has ascribed to that decision something which the division really did not determine. The report says that the aggregate resulting from an aggregate-of-intermediaries rule applied to a joint class rate was there held not to be a class rate, and, inferentially, was therefore a commodity rate within the meaning of note 4 of an intermediate rule. I discover no such finding in the report cited. The determination was that when a joint commodity rate was subject to a rule providing—*If the charges accruing under the class rates published in the following tariffs *** are lower, such lower charges will apply ****,—reference to 'the class rates published in the following tariffs' meant only the class rates actually published in those tariffs and not lower aggregates, elsewhere published resulting from applying the aggregate-of-intermediates rule in the class tariff. That conclusion was not based on the fact that the lower combination resulting from the aggregate rule was composed of commodity factors. The report cited does not even show whether the combination was composed of commodity rates or class rates. It merely took the general position that a rule under which a commodity rate alternates with a class rate should be read literally, and if the class rate itself (irrespective of any rule that would set it aside under certain circumstances) is not lower, it was not proper to go to other rates in determining whether the commodity rates should be set aside. In my opinion, the same conclusion would have been reached in that proceeding if both factors of the aggregate had been class rates.

"But, the proceeding cited negatives one of the key conclusions here made, namely, that an aggregate of separately published factors actually appearing in two other tariffs nevertheless must be held to be in the class tariff and, therefore, in a single tariff. Without that conclusion the whole report falls. But is it not clear (although not literally so stated) that the principal ground on which it was there held that the aggregate did not alternate with the joint commodity rate was because such aggregate was not 'published in the following tariffs'? In other words, in determining the charges to be alternated with those resulting from the commodity rate, only charges actually published in the class tariffs, and not charges indirectly incorporated therein as a result of an aggregate rule, were to be considered.

"The report will have the effect of overruling numerous decisions. For example, in *Rogers v. Cincinnati, N.O.&T.P. Ry. Co.,* 258 I.C.C. 553, an interpretation similar to that here made was urged by the defendants. We were not concerned with the question of joint class rates and the aggregate rule. Nevertheless, our pronouncement as to the proper construction to be accorded note 4 is pertinent here. At page 557, we stated:

> 'The note clearly renders the intermediate rule inapplicable in instances where a joint or local commodity rate is specifically provided from origin to destination on the same article in the same tariff containing the rule, or in any other tariff (but not tariffs), over the same route.'

"In *Firestone Tire & Rubber Co., of Calif. v. Southern Pac. Co.,* 243 I.C.C. 157, we said:

> 'A commodity rate to take precedence over a commodity rate from and to the same points resulting from the application of an intermediate rulé must be a specific commodity rate.'

"It certainly is inconsistent with that decision to call a sum of two rates found in different tariffs resulting from a formula in a class rate tariff 'a specific commodity rate.' Again, in *Rogers v. Cincinnati, N.O. & T.P. Ry. Co.,* supra, we said:

> 'The intermediate rule established the (Rochester) 43-cent rate to Detroit just as effectively over the route of movement as though that joint rate has been specifically named to that destination in the applicable tariff.'

"A similar finding was made in *Lustberg Nast & Co., Inc. v. New York, N.H. & H.R. Co.,*229 I.C.C. 684. The following principle was enunciated in various proceedings (*Miller & Lux, Inc. v. Southern Pac. Co.,* 102 I.C.C. 137; *Boguc Supply Co. v. Nevada Copper Belt R. Co.,* 96 I.C.C. 434; *Fruen Grain Co. v. La Crosse & S.E. Ry. Co.,* 132 I.C.C. 747):

> 'An intermediate provision established specific rates just as positively, plainly, and legally from unnamed intermediate points as if such points were specifically named.'

"Under the findings now made, the discovery of a commodity rate to a more distant point subject to an intermediate rule will not enable the user of the tariff to accept such rate as applicable, but he must

"(1) find the class rate;

"(2) see whether it is subject to an aggregate rule;

"(3) if there is one (and there nearly always is), locate the lowest aggregate over any route over which the class rate applies (and often class tariffs contain little or no specific routing, thus meaning almost any conceivable route); and

"(4) determine whether the aggregate is composed of commodity-rate factors; and

"(5) if it is so composed, apply such combination whether higher or lower, often resulting in long-and-short-haul violations of the fourth section. Thus the report will destroy the real effect of many prior reports, and likewise will substantially reduce the benefits which both shippers and carriers now derive from the use of the intermediate rule.

"I therefore dissent.

"I am authorized to state that COMMISSIONER PORTER and LEE concur in this expression."

Tariff Must Be Considered In Its Entirety

The courts have held that a cardinal rule in the construction of statutes, which may be applied to tariffs, is that effect is to be given, if possible, to the whole, to every word, clause, and sentence, and in construing a railroad tariff or other transportation agency's tariffs, as in the case of other documents, the entire instrument must be visualized, and effect must be given to every word, clause, and sentence, to the end that general and specific provisions in apparent contradiction may subsist together. *Pillsbury Flour Mills Co. v. Great Northern Ry Co.,* 25 Fed. (2d) 66; *Van Dusen Harrington Co., v. Northern Pac. Ry. Co.,* 32 Fed. (2d) 466.

Therefore, in encountering an apparently technical situation the first rule to be followed is that laid down by the courts and adopted by the Commission in the cases of *United Shoe Machinery Corp. v. Dir. Gen. and Pennsylvania Railroad Company,* 55 I.C.C. 253, 255, and *Andrew Murphy & Son, Inc., et al, v. Ann Arbor R.R. Co., et al,* 147 I.C.C. 449, 450.

In the case first referred to, the Commission said:

"The Commission cannot consider only the part of the rule that is favorable to complainant's contention and ignore other provisions of the tariff that make clear its application. *The tariff must be considered in its entirety,* considering both the limitations on its title page and rules contained therein."

In the second case, the Commission held as follows:

"In many cases we have found that the intention of the framers of a tariff is not controlling and that tariffs are to be construed according to the language used. But we have also found that a fair and reasonable construction should be given in the light of all the pertinent provisions considered together. The terms used in a tariff must be construed in the sense in which they are generally understood and accepted commercially, and neither carriers nor shippers can be permitted to urge for their own purposes a strained and unnatural construction."

In *Old Colony Furniture Co., v. B. & M. R.R.,* 270 I.C.C. 373, the Commission said:

"In construing a freight tariff or classification, the entire schedule must be scrutinized and pertinent provisions given due effect, in order that general and specific provisions in apparent contradiction may subsist together, the specific provision qualifying and supplying exceptions to those of general application."

The recognized rule of tariff interpretation, therefore, is that all pertinent provisions must be read together and all portions of the tariff given effect.

Tariffs Containing Ambiguous Provisions

It should be remembered that if there is a genuine technicality and the tariff is actually ambiguous, and not merely difficult of interpretation, the shipper will be given the benefit of the doubt. This is borne out in numerous decisions of the Commission, of which the following, as contained in the cases of *Republic of France v. Director General, as Agent,* 81 I.C.C. 174, 178, and *Sauers Milling Co. v. L. & N. R.R. Co., et al,* 216 I.C.C. 358, 361, respectively, are typical. In the Republic of France case the Commission stated:

"Whatever may have been the intention of the carriers publishing the tariffs, the rules must be interpreted according to their wording and whatever ambiguities exist must be construed against the publishing carrier."

And in the Sauers Milling Co. case, the Commission said:

"In construing tariff provisions with respect to which, as here, there is a reasonable doubt as to the meaning, we have uniformly adhered to the view that such doubt must be resolved in favor of the shipper."

It should be understood, however, that it is only in cases where the keenest traffic judgment and experience are brought to bear upon a given situation and the decision still hangs in the balance that the Commission will favor the shipper on the plea of ambiguity. There must be an equal division of the equities in the case, or the preponderance of evidence under a strict interpretation of the tariff must lean to the side of the shipper, before he should expect to win his point by an appeal to that body.

The attitude of the Commission and the courts on this question of ambiguity in tariffs is indicated in the following excerpts from their decisions:

> "The terms of tariffs should be clearly stated so that misunderstanding may be avoided, and carriers should clarify their tariffs so as to eliminate any existing confusion. *Middlewest Freight Bureau v. C.B. & Q.R.R.*, 309 I.C.C. 303."
>
> "The railroads are the drafters of tariffs which, the same as any other written instruments, should be construed strictly against them and in favor of the shipping public. *Calcium Carbonate Co. v. M.P. R.R.* 329 I.C.C. 458."
>
> "Although doubt as to the meaning of a tariff provision must ordinarily be resolved in favor of a shipper, the doubt must be a reasonable one. *Guerdon Industries v. P. R.R.* 315 I.C.C. 277.
>
> "Tariffs are written by carriers and accordingly are construed in shipper's favor in case of ambiguity. If the Commission's construction is not in accord with carrier's understanding and intent, its remedies are either a direct attack on those findings, or amendment of the tariff to correspond with its intent. *United States v. G.N. R.R.* 337 Fed. (2d) 243."

An illustration of the foregoing rule is contained in *Smoke Flue Cleaning Compounds, Transcontinental,* 268 I.C.C. 619, wherein the shipper contended that the rates applicable on smoke flue cleaning compounds are those carrying the description "Cleaning, Scouring, or Washing Compounds, N.O.S.: *** liquid *** dry *** ," listed in a tariff item captioned "Soap, Washing Compounds, and other articles as designated;" The designation "smoke flue cleaning compounds" does not appear in the commodity tariffs.

Carriers maintained that the rates on "cleaning, scouring, or washing compounds" were not applicable on smoke-flue cleaning compounds as the tariff provided rates on carbon-removing compounds.

Smoke-flue cleaning compounds, a powder, is placed on the fire in furnaces, stoves, boilers, and other heating or power-producing equipment. Its action causes the soot in flues and smokestacks to disintegrate. Dictionary definitions of soot describe it as being composed chiefly of carbon, but as also containing ash and other substances. Carbon-removing compounds are advertised as such for the removal of carbon, gums, and other adhesive material from internal-combustion engines. Originally, in 1924, carbon-removing compound was listed in the freight tariff under "Drugs, Medicines, and Chemicals." In 1931, it was published in separate tariff items but with the same rate basis. Shippers argued that since carbon is only one element constituting soot, the rates on carbon-removing compounds should not be applied on its products. The percentage of carbon in the material which forms in a motor is higher than in soot.

Considering all these facts, the Commission held, that there is a reasonable doubt as to the applicability of the higher of the rates, published for application on carbon-removing compounds, and this doubt should be resolved against the carriers. They found that the commodity rates on cleaning compounds were applicable.

Specific v. General Descriptions

Another interesting but troublesome phase of tariff interpretration involves two seemingly applicable descriptions, one specific and the other general. Here the Commission and the courts have consistently held that where a commodity shipped is included in more than one tariff designation, that which is more specific will be applicable. This principle, of course has no application where one rate is a commodity rate and the other a class rate. Commodity rates take precedence over class rates unless provision is made in the tariffs for the alternation of the class rates with the commodity rates.

Exceptions ratings, like commodity rates, take precedence over the ratings in the classification and where a broad general description is published in an exceptions tariff and a specific description in the classification the exception rating is the one to apply.

An illustration of this principle of specific v. general is found in *U.S. Industrial Alcohol v. Director General,* 68 I.C.C. 389, 392, in which the Commission said:

> "There can be but one applicable rate at the same time over the same route between the same points on the same traffic *** commodity rates take precedence over class rates and specific commodity rates take precedence over general commodity rates."

Again, in *Durez Plastics & Chemicals v. C.M.St.P. & P. R.R.,* 259 I.C.C. 335, the Commission held:

"It is not a strained or unnatural construction to hold that when the carriers publish a rate applicable to a broad and unrestricted commodity by description and an article is shipped that clearly falls within that description, the rate published in connection therewith is applicable even though a more specific description is contained in the classification."

In connection with the determination as to whether a certain article comes within the generic term, while it has been held that the generic designation takes an article out of the classification even though the classification description is more specific, it must be remembered that, necessarily, all commodities coming within a generic term are not specifically described. Therefore, if a commodity falls within a specific description, the rating for that specific description is applicable, and not an "n.o.i.b.n." rating. See *Darling & Co. v. New York, C. & St.L.R. Co.,* 213 I.C.C. 418, 420, 421.

Use to Which an Article is Put and Description For Sale Purposes

The Commission has, in many cases, ruled that "it is the character of an article from a transportation standpoint, and not the use to which it is put, that determines the applicable rate or rating," Service Pipe Line Co. v. C. & N.W. Ry., 301 I.C.C. 545, 548. The use of an article, however, may be considered in determining what an article actually is.

In *Hyman-Michaels v. C.R.I. & P. R.R.,* 308 I.C.C. 339, the Commission said:

"Nature and character of a commodity at time it is tendered for shipment determines its identity for transportation purposes. Facts weighing heavily in making that determination are producer's description of the commodity for sales purposes, its value and general use, its billing, and how it is regarded in the trade."

The application of these principles is given in *Spear Mills, Inc., v. Alton R. Co. et al,* 268 I.C.C. 330, wherein the shipper sought determination of the applicable rate on a carload of ground bones shipped from Chicago, Ill., to Joplin, Mo. Charges were assessed on the basis of a 28 cent rate. Carriers sought to collect undercharges to the basis of the class D rate of 40 cents and shipper opposed payment of undercharges in excess of 31 cents.

The commodity shipped was made from animal bones which had been ground, after having been cooked under steam at high temperature for several hours in order to free the glue liquors, grease and animal tissues, and to make the product sterile so far as any germs or bacteria were concerned. This commodity was not a complete food, but was mixed with other ingredients for the purpose of adding the necessary elements required in a balanced feed for livestock and poultry. The shipment was billed as "600 bags bone meal feed." The three pertinent tariff items in issue were as follows:

Western Classification:

Feed, Animal or Poultry:
Blood flour, blood, bone or meat meal;
feeding tankage or dried meat scraps: *** .... Class D (22 ½% of first class)
Bones, n.o.i.b.n., ground or not ground, *** .... Class D (22 ½% of first class)

Exceptions to Western Classification:

Bones (other than human or fresh meat
bones), ground or not ground .... Column 17 ½ (17 ½% of first class)

Shipper contended that the commodity shipped was simply ground animal bones, that nothing had been added, and that bones, whether aged and cleaned of all animal matter through exposure to the elements for a long period of time, or cleaned of such matter by a steaming process in a few hours, were still bones regardless of the fact that they were described as bone meal for the purposes of sale, and were billed as bonemeal feed. Carriers claimed that this commodity was not merely ground bones, but was a bonemeal feed made from selected bones which were subjected to treatment to remove certain ingredients, and the residue, not the original bones, was ground sufficiently fine to constitute a meal. Complainant's contention in that respect was supported by an affidavit by the shipper's traffic manager, that while the shipment was billed as bonemeal feed, as a matter of fact it actually consisted of ground bones, other than human or fresh meat bones.

Inasmuch as the commodity was not a complete food, but was used only to supplement the diet of livestock and poultry, complainant contended it was not a feed anymore than is salt, various minerals, linseed oil for dosing, or other commodities which are fed in small quantities to livestock and poultry. To refute complainant's argument that the commodity was not a complete animal or poultry food and therefore should not be considered as a feed, carriers pointed out that it was doubtful whether there was any feed, other than a mixture which would of itself provide a complete or balanced feed for livestock or poultry.

In support of their contention that the class D rate of 40 cents was the applicable rate on the commodity, carriers relied chiefly on the principle that a description placed upon a commodity by the manufacturer thereof for the purpose of sale, also fixes the identity of the commodity for transportation purposes. On the other hand, complainant argued that this principle may not be applied against a third party, because it is in the nature of a declaration against interest. The Commission said complainant's position was well taken. The Commission has stated repeatedly that a shipper is not bound necessarily by commodity descriptions contained in invoices or billing. This is particularly true of statements contained in invoices or billing which are made by someone other than the complainant in a proceeding.

Although the Commission has concluded, in certain proceedings, that a manufacturer's statements may properly be considered in determining the nature of a commodity for classification purposes, it also has said that those statements are not controlling and that the rates applicable are those which the governing tariffs provide on the commodity actually shipped.

Complainant contended also that inasmuch as the lower rating applicable on ground bones was an exception rating, it should take precedence over the classification rating on bonemeal feed. It also contended that the description in the exception tariff was more specific than that in the classification, which includes bonemeal feed along with various other feeds. The Commission said:

> "These contentions are sound. It is apparent that all ground bone is not necessarily bonemeal feed, and in construing tariff provisions with respect to which there is a reasonable doubt as to meaning, the doubt must be resolved against the maker. Moreover, it is well settled that where two commodity descriptions are published, both of which might apply to the commodity shipped, the shipper is entitled to the one taking the lower rate.
>
> "It is found that the rate of 40 cents sought to be charged on the shipment under consideration was inapplicable, and that the 31-cent rate on ground bones was applicable."

Conflicting Rates

It is a well accepted principle of tariff interpretation that when two tariffs are equally appropriate the one which provides the shipper with the lower rate is to be applied. This is consistent with what we have encountered in connection with ambiguous provisions in carrier tariffs, i.e., that carriers are responsible for their published tariff provisions and where conflicting provisions are maintained, those resulting in the lowest charge to the shipper are applicable. See *Farmers Union Grain Term. Ass'n. v. Can. Nat'l. Ry.,* 315 I.C.C. 559.

In *Gar Wood Industries v. A. & S. R.R.,* 263 I.C.C. 611 the Commission ruled:

> "Where tariffs contain conflicting commodity descriptions and rates which are equally applicable to the commodity shipped and neither the tariff nor the classification indicate a contrary intention, the lower of the two conflicting rates is applicable. In other words, where two descriptions are equally appropriate, considering all provisions of the governing tariffs and classifications the shipper is entitled to the benefit of the lower rate. However, where one of the tariff descriptions is more specific, the rate published in connection with such description should be applied to the exclusion of a rate published in connection with a general commodity description, even though the latter carries a lower rate."

Tariff Errors

Considering the tremendous volume of tariff matter published each year by the carriers, it is inevitable that some errors will be made. When they are discovered, the tariff is usually corrected promptly by the issuance of a supplement. However, in those cases where the error is not found until after many shipments have moved, a problem arises which is much more difficult than the mere correction of the tariff. If the published but erroneous rate is higher than it should be, the shipper will, naturally, wish to recover the difference between the rate charged and the rate which should have been published and charged. This, however, is not an overcharge and recovery can not be had through the usual overcharge claim channels. The incorrect rate is the published rate and an error in tariff publication is no justification for a departure from the published rate which, according to Section 10761, is the only rate that may be charged.

In situations such as this, the shipper may ask the carrier to file a

Special Docket Application with the I.C.C. requesting authority to refund the difference between the rate charged and the rate which should have been published and charged. If the carrier refuses to take such action, shipper should file complaint with the I.C.C. In any event, one way or the other, the shipper can recover.

If the situations were reversed and the erroneous rate was lower than it should have been, the carrier, unfortunately, is without recourse. Under the Act the carrier must charge the tariff rate and there is no provision in the law enabling the carrier to take any action against the shipper. Here we have another case of the carrier being held responsible for its own actions. Tariff errors must be construed against the framer of the tariff.

In *Tobin Packing Co. v. B. & O. R.R.*, 299 I.C.C. 221, the Commission said:

> "Error in publication of rates does not relieve carriers from their duty under Section 6(7) to charge rates specified in the tariffs in effect on date of shipment. Proof of such error does not justify departure from published rates, even though shippers have full knowledge that rates were published by mistake. The Commission has no authority to permit or require correction of tariff error nunc pro tunc (now for then)."

In analyzing these findings of the Commission and applying them to the feature of technical interpretations, we observe a recognition of the principle in traffic that a specific item always supersedes a general provision, and that where a commodity shipped is included in more than one tariff designation, that which is more specific will be held applicable. It will be observed that it is the duty of the carriers under the law to plainly state their transportation charges so there will be no conflict or ambiguities, and if they fail to do so, then they should bear the burden of their own short-comings by being compelled to assess the lower of the two equally appropriate rates—even though such lower rate may not have been the one intended to be applied by the framers of the tariff. Furthermore, when errors are made in the publication of rates, the carrier must bear the responsibility for such errors.

CHAPTER 17

ROUTING AND MISROUTING

General

Section 10763 of P.L. 95-473 gives shippers the right to designate in writing the route via which their property shall be transported to destination via rail carriers and makes it the duty of the carriers involved to transport the property according to the routing instructions specified in the bill of lading.

With regard to the publishing of routing instructions, Rule 4(k) of Tariff Circular 20 provides that tariffs shall contain "routing over which the rates apply, stated in such manner that such routes may be definitely ascertained." This rule continues by stating, "this must be one of the following plans: (1) by providing that the rates in the tariff apply only via the routes specifically shown therein, or (2) by providing that the rates apply via all routes made by use of the lines of the carriers parties to the tariff except as otherwise specifically provided in the tariff."

Rule 4(k) also provides that "In lieu of showing in rate tariffs the affirmative routes provided in plans (1) and/or (2) above, such affirmative routes may be published in a separate publication (or publications) filed either by a carrier or by an agent, and specific I.C.C. reference must be made in the rate tariff to such guide."

There is nothing in P.L. 95-473 that gives shippers the right to route shipments via motor carriers, water carriers or freight forwarders. However, Tariff Circular MF No. 5 (common carriers by motor vehicle) provides in Rule 8(a):

> "Tariffs containing joint rates shall specify routes over which such rates apply or refer to a separate tariff or tariffs for such provisions. The routes must be stated in such manner that they may be definitely ascertained."

In view of the provision of Tariff Circular MF No. 5, it is apparent that despite the absence of a specific authorization in P.L. 95-473 shippers may, by virtue of the routing provisions contained in motor carrier tariffs, indicate in the bill of lading the route they wish shipment to travel. Similarly, the motor carrier, while not specifically required by law to observe the routing instructions of the shipper, is obliged to follow the routing contained in its tariffs and must also comply with those provisions of the law requiring the establishment of reasonable rules, regulations and practices.

Routing Contained In Tariffs

When in conformity with the provisions of Rule 4(k) of Tariff Circular 20, the rail carriers provide in a tariff that "the rates apply via all routes made by use of the lines of the carriers parties to the tariff" the rates must be applied strictly in conformity with that provision. In *Sioux City Foundry and Boiler Co. v. C.B. & Q. R.R.*, 318 I.C.C. 13, the Commission stated:

> "If a carrier desires to limit the routing of traffic to specific lines of its system, it is free to do so. However, the failure to provide internal routing does not prevent the use of an open route which is not unduly circuitous."

Again, in *Producers Cooperative Oil Mill v. St.L.S.F.*, 322 I.C.C. 267, they held:

> "Where a tariff specifically provides for the application of rates over all routes composed of lines of carriers parties thereto, the burden is on the carriers to publish definite routing restrictions if they deem any routes to be in any way unreasonable."

Where the tariff naming the rate contains specific routing instructions or refers to a routing guide containing specific routes, the rates are of course applicable only via the routes shown. Here again, however, it is the duty of the carrier to place specific restrictions in the tariff where it wishes to have traffic move via only one of several available lines or branches. In *Ark. Farmers Plant Food Co. v. A.C.L. R.R.*, 296 I.C.C. 289, the Commission said:

> "When a carrier's tariff does not specify over which route of the carrier a rate applies and a longer available route is not excessively circuitous, the rate must be regarded as applying over both, since the tariff could have been so drawn as to preclude application of the considered rate over the indirect route, but did not do so."

Where there is involved the question of local routing, i.e., the movement via certain junctions or diverging lines of a particular railroad participating in the through route the Commission has held in many cases that shippers cannot be charged with knowledge of matters of operating conditions regarding the handling of through traffic and that it is the responsibility of the carrier to provide appropriate routing restrictions.

Although a shipper has the privilege of specifying routing via any and all carriers parties to the tariff, there is one feature in connection with "open routing" that should be borne in mind. Where there is a choice of two or more routes, a carrier cannot be forced to haul a shipment via a route which would result in a violation of the Act. The U.S. Supreme Court brought this out very clearly in *Great Northern Ry. v. Delmar Co.*, 283 U.S. 686, in the following language:

> "The pertinent facts are that numerous shipments of grain originated at points on the line of the railway in Minnesota, North Dakota and South Dakota. They were originally billed to Minneapolis. After arrival there they were reconsigned by the Delmar Company, in the same cars, to Superior, Wisconsin, where delivery was made. The entire movement from the points of origin to Superior was over the rails of the Great Northern. The shorter route from the places of shipment to Superior is via Willmar. The longer, which the cars in question traveled,—via Minneapolis, involves passage through the congested railroad terminals in that city, with incident traffic difficulties and delays not encountered on the more direct one. The difference between the two in mileage from the shipping points to Superior varies from 12 to 23 per cent. The carrier collected its local rates from origin points to Minneapolis, plus a proportional rate of 6.5 cents beyond. The combination of rates so exacted were higher than the through rates specified in the tariffs for the transportation of grain from these points to Superior.
>
> " ... The Commission sustained the contention of the Delmar Company that the quoted through rates from points of origin to Superior applied to shipments routed via Minneapolis, since the tariff did not expressly restrict their application to the shorter and more direct route via Willmar. This finding was the basis of the award of reparation.
>
> " ... The railway maintains that in the circumstances here presented the tariff may not be so construed as to render the specified through rate applicable to shipments by way of Minneapolis. This would be contrary to established custom and would occasion violation of the long-and-short haul clause of the Interstate Commerce Act.
>
> " ... The railway can transport the shipments over the shorter and customary route without violating Section 4; but if the tariff is construed to require it to take them over the longer route, it must violate that section and incur the resulting penalties. In this situation, we think the tariff should be construed as applying only to the shorter route, and not as giving the shipper the option between the two routes at the through rate. This conclusion is in accord with the principle that where two constructions of a written contract are possible, preference will be given to that which does not result in violation of law."

Routing contained in motor carrier tariffs is not appreciably different from that found in rail tariffs. Basically the routing, whether in the rate tariff itself or a special routing tariff indicates the points at which the various carriers, parties to the tariff, will interchange freight. Once the through route has been voluntarily established by the carriers and published in their tariff it must be observed. It must be remembered, however, that motor carriers are not required to establish through routes and joint rates. Section 10703(a)(4)(A) is permissive and not mandatory, it clearly states:

> "A motor common carrier of property may establish through routes and joint rates and classifications applicable to them with other carriers of the same type, with rail and express carriers, and with water common carriers, including those referred to in subparagraph (D) of this paragraph."

In *Murray Co. of Texas v. Morrow, Inc.*, 54 M.C.C. 442, the Commission ruled:

> " ... while a motor carrier must under Section 217(b), observe its tariff rates over the route of movement, it also must observe just and reasonable routing practices. This duty is not affected by absence from Part II of any specific grant to shippers of the right to designate routes by which their property shall be transported."

Unrouted Shipments

Very frequently shipments come into the carriers possession with no routing specified on the bill of lading. In some instances this may be due to lack of knowledge of available carrier routes or may be the desire of the shipper to place the responsibility for properly routing the shipment on the carrier. In any event, when this occurs, the originating carrier may either select a route for the through movement to the ultimate destination or forward the shipment to a point on its line where it can be delivered to a second carrier operating to, or in the direction of, the final destination.

Assuming there are several routes available, all of which involve one or more additional carriers, how should the originating carrier decide? If the rates are the same via all routes, there is no problem; the originating carrier may select any one of the routes. In all probability, it will be the one which gives the originating carrier its longest haul. But what if the rates differ? In those situations where there is a choice of routes and the applicable rates vary, the Commission and the courts have ruled that the carrier has the obligation to forward the shipment via the cheapest available route.

Surprisingly enough, the specific and unambiguous statement, "via the cheapest available route," has created some problems requiring clarification by the Commission. One of the first questions is, must the carrier to whom the shipment was tendered by the shipper turn the shipment over to a competing carrier participating in a lower rated route. The I.C.C. has ruled on this in many cases and in *Hope Brick Works v. A.T. & S.F. Ry.*, 200 I.C.C. 215, they said:

> "An originating carrier is under no obligation to deliver shipments to its competitor at point of origin even though the competitor participates in a cheaper route to the destination of the particular shipment, as the originating carrier is entitled to a line haul."

The fact that the originating carrier is entitled to a line haul does not necessarily mean the longest possible haul. In resolving this problem, the Commission, in *Deisel Wemmer-Gilbert v. P.R.R.*, 218 I.C.C. 137, said:

> "Where a shipment is tendered to an originating carrier unrouted and the carrier has several routes over which the shipment could have been forwarded, the lowest rated route being a 3-line route, the carrier is obligated to forward the shipment over such 3-line route rather than over its own higher-rated single line route, since the shipper could have directed routing over the lowest-rated, in which event his instructions could not have been lawfully ignored."

In the *Murray* case, 54 M.C.C. 442, previously referred to, the Commission had before it the question of unrouted shipments via motor carrier and ruled as follows:

> "When no routing instructions are given, a motor carrier has the duty to select the least expensive route unless it is an unreasonable one.
>
> "Complainant has shown that defendant maintained different rates on the same traffic over more than two open routes, and the burden was cast upon defendant to rebut the presumption that it acted unreasonably in transporting the unrouted shipments over the higher-rated route.
>
> "Lower-rated routes cannot be condemned as unreasonable merely because their use would have resulted in a three-line haul. Neither can the use of the higher-rated route be justified on the ground that the defendants operating practices forbid its equipment to be transported by more than one connecting carrier, nor because its equipment would have arrived at a point of interchange too late to permit transfer of lading to a connecting carrier. And the fact that defendant would have been short hauled had it selected one of the lower rated routes, was of its own making by its voluntary participation in those routes."

Routing Specified By Shipper

Under Section 10763 of P.L. 95-473, the shipper has the right to designate in writing the rail route via which his property shall be transported. These instructions are usually inserted on the bill of lading and, again in accordance with the provisions of Section 10763:

"A carrier directed to route property transported under paragraph (1) of this subsection must issue a through bill of lading containing the routing instructions and transport the property according to the instructions. When the property is delivered to a connecting carrier, that carrier must also receive and transport it according to the routing instructions and deliver it to the next succeeding carrier or consignee according to the instructions."

The provisions of the statute are certainly explicit and there can be no question as to the rights of shippers and the duties of carriers where the routing of freight via rail carriers is concerned. The problems arise in the application of these provisions.

In the exercise of their right to route their shipments, shippers frequently create, perhaps unknowingly, problems which result in higher charges than would have applied had different routing instructions been given. It must be remembered that the carrier is required by law to follow the shipper's routing instructions and when the shipper specifies a route on the bill of lading and no rate is shown, the carrier cannot be held responsible if the rate applicable over the route shipment traveled is higher than the rate over some other route. (See *Midwest Industries v. B. & O.,* 303 I.C.C. 319.)

Where the shipper specifies the junctions or points of interchange in addition to the carriers, there can be no doubt as to exactly how shipment should move and the carrier will be governed accordingly. However, if no junctions are indicated, the choice is left to the carrier. In situations such as this, the Commission has held:

"When shipper specifies the route but does not indicate junctions or points of interchange through which the shipment is to move and no rate is shown, the carrier has the duty of selecting the route over which the lowest rate applies. *Boyertown Burial Casket v. D.L. & W. R.R.,* 305 I.C.C. 333."

Similarly, when a shipper specifies partial routing only, it is the carrier's duty to forward the shipment over the cheapest reasonable available route consistent with the partial routing instructions which afford the originating carrier a line haul. (See *Adel Canning & Pickling Co. v. G. & F. R.R.,* 287 I.C.C. 239.)

Another illustration of partial routing, which is actually no routing at all, is where the shipper indicates on the bill of lading only the desired delivering carrier. In *Duluth Chamber of Commerce v. C.B. & Q. R.R.,* 210 I.C.C. 652, the Commission said:

"If shipper specifies the delivering carrier only in his routing instruction, it is the duty of the carriers to forward the shipment via the cheapest route consistent with such instructions."

Also on the same general subject of delivering carrier, the Commission has said that carriers are not presumed to know where consignees desire delivery and if through failure of the shipper to state the delivery desired, a shipment arrives on the line of a carrier other than the one on which the consignee is located, or the one most convenient to his plant, carriers may not be charged with misrouting. *Middleton, Inc. v. N.S. R.R.*, 215 I.C.C. 411.

As an added precaution, some shippers indicate on the bill of lading both the route they want shipment to follow and the applicable rate. If an error is made and the rate does not apply via the route indicated, a conflict exists and in such situations it is the duty of the originating carrier to ascertain from the shipper whether the routing specified should be observed or whether the shipment should be forwarded over a route which would protect the rate shown. Failure to perform their duty renders the originating carrier liable for any additional charges which may result from such misrouting. (See *Westinghouse Electric Corp. v. C.G.W. Ry.*, 299 I.C.C. 527; *Empire Oil & Refining Co. v. A.T. & S.F. Ry.*, 205 I.C.C. 239 and many others.)

In *Johns Manville Products Corp. v. Alton R.R.*, 269 I.C.C. 739, the Commission had before it the problem of a shipment from Waukegan, Ill. to Boise, Ida. with a stop-off for partial unloading at Pueblo, Colo. Shipment was routed by the shipper. The agent of the originating carrier did not bring to the attention of the shipper the fact that the rate of 110 cents, also shown in the bill of lading, did not apply via the route specified. The Commission declined to find that shipment had been misrouted, stating:

"In numerous cases, the Commission has found that, where there is a conflict between the rate and the route shown in a bill of lading, it is the duty of the carrier's agent to direct the shipper's attention to the conflict, and the failure to do so renders the carrier liable for any resulting damage. Here, however, the complainant also requested stop-off for unloading. As the shipment moved as routed, over the lowest-rated available route, permitting the stop-off, the shippment ws not misrouted."

With regard to shipper's right to route shipments via motor carriers, the Commission in *Eastern Aircraft v. Olson & Son*, 47 M.C.C. 363, held:

"Part II of the Act contains no provisions which authorize shippers to specify routing. Prior to the Mann-Elkins Act of June 18, 1910, now embodied in Section 15(8), Part I of the Act likewise contained no provisions which specifically authorized shippers to designate routing, but in a number of proceedings, under statutory provisions somewhat similar to Section 216(c) of the Act, the Commission then found that it was the duty of the initial carrier to obey the routing instructions specified by the shipper, and that its failure to do so made it liable for misrouting."

Routing From or To Unnamed Intermediate Points

Not infrequently, the point of origin or point of destination of the shipment is not specifically named in the tariff containing the through rate and the use of an intermediate clause is authorized. If routing is not provided in the tariff, then, as pointed out above, the rates properly may be applied over all reasonably available routes of the participating carriers. In the application of this principle, the question arises—what is a reasonable route?

There is no definite rule of law fixing the degree of circuity from or to which intermediate rates may or may not be applied. In no case has the Commission fixed a standard by which to determine what constitutes a natural or an excessive route. Questions of this kind necessarily must be determined in the light of the facts and circumstances surrounding the particular movement.

Generally in situations of this nature, two questions are presented for consideration, viz: is the route in question a reasonable and practicable route over which the shipper has a right to require his shipment to move, and has the shipper the right to specify local routing over the lines of any carrier? The Commission has held that it is a well-settled general principle that the intermediate rule applies over a circuitous route provided that the circuity is not excessive. (*Wemmer-Gilbert Corp. v. Pennsylvania R. Co.,* 218 I.C.C. 137.)

It is also an established principle that the intermediate rule does not apply over a route that is unduly circuitous or that is regarded as impracticable, illogical, or unnatural because of other operating difficulties, even though the routing is not in terms restricted. (*G. Caruso & Co. v. St. Louis-S.F. Ry. Co.,* 156 I.C.C. 429, and cases therein cited.)

In the many cases before the Commission involving this subject, no fixed rule has been observed in regard to the degree of circuity or other operating difficulty considered as rendering a route unavailable. For example, in *Miner Lbr. Co. v. Pennsylvania R. Co.,* 161 I.C.C. 801, the Commission said:

"In no case where the application of Rule 77 has been considered have we established any definitely settled rule of what is or is not a reasonable or natural route. Every case has been decided on its own merits."

In *United States v. Southern Ry. Co.,* 142 F. supp. 741, 742, the district court held that a "route" for intermediate-point rate purposes is subject to the limitation that it must not be unreasonable; unreasonableness depends upon extent and direction of circuity and the commercial usage, of a route. The Commission's finding that as there was 22.5 percent circuity and route was not in commercial usage, intermediate rule was inapplicable, sustained.

An interesting case involving through routes and the use of an intermediate clause is *A.E. West Petroleum Co. v. A.T. & S.F. Ry.,* 212 Fed. 2nd 812 (1954). The facts were as follows: lubricating oil was shipped from Bradford, Pa., to Kansas City, Mo. A specific through rate of 67 cents was published from Bradford to Kansas City in one tariff while another tariff contained a rate of 53 cents from Bradford to points in Iowa such as Moravia and Cedar Rapids which tariff was not restricted as to routing and applied "via all routes made by use of the lines parties to the tariff." Since there were no restrictions prohibiting the routing of freight to the Iowa destinations via Kansas City and as the tariff contained an intermediate point rule permitting the application of the rate to the next more distant station at the unnamed intermediate point, the shipper contended that the 53 cent rate should be applied on the shipment to Kansas City. The I.C.C. found for the complainant and awarded reparation. The Federal District Court held that the I.C.C. decision was unsound in law and unreasonable.

The Court of Appeals upheld the decision of the District Court and in their opinion stated:

"The crucial dispute is whether Kansas City is, for freight rate purposes, an 'intermediate point' between Bradford and the Iowa stations within the meaning of certain general provisions of the Jones Tariff. These provisions concern intermediate point rates and routing instructions.

"The view of the Commission seems based upon the conception that the routing provisions should be given the unlimited meaning under its interpretation for two reasons: (1) because the carrier can specifically provide that the tariff applies only via the routes shown therein; and (2) that the routing over which the tariff rates apply must be so stated as to be definitely ascertainable by the shipper. Each of these propositions is worth examination.

"(1) The routing provision does give the carrier the right to specify therein the routes either by express designation or by stated exceptions to an 'all routes' situation. The railways of the country form an intricate system of lines with innumerable physical connections so that it is possible to form almost countless combinations of lines (or segments thereof) into routes over which a car might be continuously carried. Each of these lines has its individual problems in setting up rate structures which will provide operating revenue and yet fit into all sorts of economic, competitive, physical and various other practical conditions. The rate structure of a line must also be synchronized into the rate structures of all of the railway systems with which it physically connects. The complexities of this broad situation are almost beyond enumeration.

"There are practical difficulties to be encountered when an intermediate-point rate and routing situations are involved. Few things can be more practical and, therefore, more factual than railway rates. The fantastic and utterly unreal situations which are worked out by traffic experts for shippers can hardly be more clearly revealed than placing a railway map upon the factual situation shown in this case and in the *Pure Oil and Hermann-Brownlow* cases.

"By Rules of the Commission governing publication of freight tariffs (Rule 4(k) Circular No. 20 and Rule 27 in Supp. 5 thereto), the routing provisions involved here are *required* to be published as part of rate tariffs applicable to intermediate point-rates. When the expert ingenuity in creating imaginary routes (as revealed in this case) and the complexity of our vast railway systems and the nature of rate structures are considered, the extreme difficulty of any carrier complying with these Rules, as construed by the Commission, without danger to itself is clear—it would mean anticipation and specific exception as to every imaginable so-called 'route.' A definite ever present peril is that it could never know, until too late, if it had underestimated the superior skill of the shipper's expert in devising a route no one else had dreamed of. Obviously, this would defeat the entire purpose and usefulness of the intermediate-point rate provisions in tariffs—to lessen the intricacy and bulk of rate tariffs.

"(2) That any rate should be as clearly stated as may be for the information of the shipper is true. Also, it is true that ambiguities may be taken advantage of by him. However, these rules of construction leave untouched the other rules of construction that all pertinent parts of a tariff must be construed together and that no unreasonable or unjust meaning will be accepted—particularly where such produces absurd and obviously unintended on non-understood results. A tariff is no different from any other contract, in that its true application must sometimes be determined by the factual situation upon which it is sought to be impressed.

"Our conclusions are as follows:

"(1) Where (as here) the facts are not in dispute and the pertinent terms in a tariff are used in their ordinary meaning, the Courts have power to construe the tariff;

"(2) that general doctrines of construction of contracts are applicable;

"(3) that the 'routing' and the 'intermediate-point' rate provisions of the Jones Tariff must be so construed as to effectuate the purposes of both;

"(4) that a 'route' for 'intermediate-point' rate purposes is subject to the limitation that it must not be unreasonable;

"(5) that 'unreasonableness' depends in this case upon (a) the extent and direction of circuity and (b) the commercial usage of a 'route;'

"(6) that the undisputed facts in this case convince that a route from Bradford through Kansas City as an intermediate-point to these Iowa stations is unreasonable and, therefore, is not a 'route' within the meaning of the Jones Tariff governing 'routes' and 'intermediate-point' rates.

"Such being our conclusions, the result is that the judgment of the District Court is affirmed."

The problems which have arisen as a result of applying the intermediate point rule in tariffs which provide that rates apply over the lines of all participating carriers should no longer arise as a result of the changes in Rule 27 of Tariff Circular 20. Supplement No. 8 to the Circular reads:

"(a) Effective on and after June 10, 1959, an intermediate point rule may not be published so as to result in establishing from (or to, as the case may be) an intermediate point, a rate from (or to) a more distant point unless the tariff contains specific routing instructions showing definitely in accordance with Plan (1) of Rule 4(k) the routes through the intermediate point over which the rate from (or to) the more distant point applies."

Routing Not Dependent Upon Established Divisions

The fact that the participating carriers have not agreed to divisions cannot make a through rate inapplicable via a particular junction. To charge a shipper with such knowledge would in fact nullify the effect of the tariff. Where through joint rates are established and no routing limitations are incorporated in connection with such through joint rates, the rates will apply via all junction points and their application is not limited to only those gateways via which the participating carriers have agreed divisions. In *George C. Holt and Benjamin B. Odell, as receivers of Aetna Explosives Co., et al. v. Chicago & E.I.R. Co.*, 52 I.C.C. 393, 396, the Commission made the following statement:

"The only obstacle to the use of these rates lies in the fact that no divisions have been established by the carriers, but failure to establish divisions does not render the rates inapplicable."

In I. & S. Docket 2063, *Routing on Paper from Michigan to Illinois and Wisconsin points,* 91 I.C.C. 452-53, the Commission said:

"We have repeatedly condemned attempts to make the application of rates dependent upon the existence of divisions arrangements."

Unpublished Operating Schedules

Unpublished operating schedules have no bearing on the application of rates and in those cases where rates are named without definitely publishing routing instructions, the shipper is not bound by such unpublished operating schedules.

The shipper is charged only with the knowledge contained in tariffs lawfully on file with the Interstate Commerce Commission. Most railroads, particularly the larger ones, have standard methods of hauling and routing shipments and such methods are published by the operating departments but are not filed with the Interstate Commerce Commission, nor are they made public. In referring to such schedules, the Commission, in *Lustberg Nast & Co., Inc. v. New York, New Haven and Hartford Railroad Co., et al,* 229 I.C.C. 684, 686, said:

"Since these schedules are extraneous to the tariff containing the Intermediate Rule, the shipper is not required to take cognizance of them to determine the applicable rate."

Adjustment of Claims for Damages Resulting from Misrouting

On July 19, 1963 the Interstate Commerce Commission, in Ex Parte 235, issued an interpretative opinion concerning adjustment by carriers subject to Parts I (rail) and II (motor carriers) of the Interstate Commerce Act of claims resulting from misrouting on shipments by such carriers. The following is the complete text of their report.

"*REPORT OF THE COMMISSION*

"BY THE COMMISSION:

"This report is issued for the purpose of making public an interpretative opinion, as contemplated by the Administrative Procedure Act, 5 U.S.C. § 1002(a)(3), concerning adjustments by carriers subject to parts I and II of the Interstate Commerce Act of claims resulting from misrouting of shipments by such carriers, in order to clarify any misunderstandings resulting from the decision of the Supreme Court in *T.I.M.E., Inc. v. United States,* 359 U.S. 464, hereinafter called T.I.M.E., and to direct the attention of carriers and the shipping public to the decision of the same Court in *Hewitt-Robins, Inc. v. Eastern Freight-Ways, Inc.,* 371 U.S. 84, hereinafter called Hewitt-Robins.

"In T.I.M.E. the Supreme Court stated that a shipper by motor carrier could not maintain a post-shipment action for the recovery of damages, based upon unreasonableness, arising out of the assessment of the applicable rate. The opinion of the Court was predicated upon a statutory intent of Congress to eliminate a common-law right of action and our lack of authority under Part II of the act to award reparation. The Court also pointed out that the shipper, within the statutory period provided by the act, could assail the publication of proposed rates and thus had an adequate remedy. However, the act significantly curtailed the pre-existing rights of the carrier to set its own rates.

"Following the aforesaid decision, the Court rendered its decision in Hewitt-Robins, limited to the question of whether a shipper by motor carrier could recover damages for misrouting. The motor carrier therein assessed the applicable rate, on an unrouted shipment, over its interstate route. The Court, in distinguishing T.I.M.E., stated that in these circumstances the act did not preclude a post-shipment court action by the shipper. Unlike the situation with respect to ratemaking, there is no statutory procedure by which a shipper may challenge a routing practice prior to the movement. Furthermore, a misrouting claim would not jeopardize the stability of tariffs or certificated routes, and a denial of any remedy to the shipper would be inconsistent with the purposes of the act.

"The matters involved in misrouting cover issues inherent in the movement of property by carriers subject to parts I and II of the act. In these circumstances, we think it will be to the interest of all persons concerned if we issue a report with an appropriate order embodying our opinion.

"Notice of this interpretative opinion shall be given to the general public by depositing a copy hereof in the office of the Secretary of the Commission at Washington, D.C., and by publishing in the Federal Register.

> 'COMMISSIONER FREAS, whom Commissioner TUGGLE joins, dissenting:
>
> 'I do not reach the merits herein since, in my opinion, there is no necessity to issue an interpretative opinion as to the holding of the Supreme Court in Hewitt-Robins. Whatever doubt may have existed as to the effect of the T.I.M.E. decision on misrouting practices of motor carriers Hewitt-Robins put to rest. The latter decision furnishes ample guide to all concerned as to the handling of misrouting matters.'

"*ORDER*

"At a General Session of the INTERSTATE COMMERCE COMMISSION, held at its office in Washington, D.C., on the 19th day of July, A.D. 1963.

"EX PARTE NO. 235

"ADJUSTMENT OF CLAIMS FOR DAMAGES RESULTING FROM MISROUTING

"There being under consideration the matter of the issuance of an interpretative opinion concerning adjustment by carriers subject to parts I and II of the Interstate Commerce Act of claims resulting from misrouting of shipments by such carriers, in order to clarify any misunderstanding resulting from the decision of the Supreme Court of the United States in *T.I.M.E., Inc. v. United States,* 359 U.S. 464, and to direct the attention of carriers and the shipping public to the decision of the Supreme Court in *Hewitt-Robins, Inc. v. Eastern Freight-Ways, Inc.,* 371 U.S. 84, that a shipper by motor carrier may maintain an action in court to recover damages resulting from misrouting, the following opinion is issued:

"TITLE 49—TRANSPORTATION
CHAPTER I—INTERSTATE COMMERCE COMMISSION
SUBCHAPTER A—GENERAL RULES AND REGULATIONS
PART 1004—ADMINISTRATIVE INTERPRETATIONS

"Sec. 1004.1—Authorization for adjustment of claims for damages resulting from misrouting.—While the Interstate Commerce Commission has no authority to award damages against motor carriers subject to part II of the Interstate Commerce Act, it is of the opinion that motor carriers should adjust claims caused by misrouting; and carriers subject to part I of the Act, with respect to which the Commission does have jurisdiction to award damages, should continue to adjust claims caused by misrouting.

"(a) Applicability.—This interpretation applies only to cases in which carriers, without concurrence of the shipper, bill or actually forward or divert a shipment over a route that is more expensive than that directed by the shipper, or one more expensive than a route available (over carriers of the same type or mode as the initial carrier) in the absence of routing instructions by the shipper. This interpretation must not be used as a means or device by which to evade or defeat tariff rates or to meet the rate of a competing carrier or route, nor to relieve a shipper from responsibility for his own routing instructions.

"(b) Refund.—If a carrier misroutes a shipment and thus causes extra expense to the shipper over and above the legal charges applicable over the route that should have been used, and responsibility for its error is admitted by the carrier, such carrier should adjust the charge on such shipment by refunding to the shipper the difference in the legal charges over the route over which the shipment moved and what would have been the legal charges on the same shipment at the same time over the lower rated available route which should have been used. Such refund must be paid in full by the carrier which caused the misrouting, and must not be shared in by or divided with any other carrier.

"(c) Bearing of charges.—The prerequisites to any refund under this interpretation are admission by the carrier of responsibility for its error in misrouting the shipment, and its willingness to bear the extra expense so caused, without recourse upon any other carrier for any part thereof. Therefore, if the misrouting is discovered before the shipment has been delivered to consignee or before charges demanded for the shipment have been paid, the carrier acknowledging responsibility for the error should authorize the delivering carrier to deliver shipment upon payment of the charges that would have applied but for the misrouting and to bill it for the extra charge resulting from the misrouting; and if the misrouting is discovered after the shipment has been delivered upon payment of the charges applicable over the lower rated route that should have been used, the carrier that acknowledges responsibility for the error should pay the carrier that delivered the shipment the extra charge resulting from the misrouting, without further handling with the shipper.

"(d) Declaratory determination of lawfulness.—Where a carrier claims justification for forwarding a shipment over the higher rated route, the Commission, at its discretion, and upon appropriate petition, may make a declaratory determination of the lawfulness thereof. With respect to motor transportation, such declaratory determinations will serve only as aids to the courts or as opinions by the Commission as to the lawfulness of any adjustment or refund.

"(e) Time limitations.—Damages caused by misrouting are not overcharges. Therefore, adjustments of claims for misrouting against carriers subject to the provisions of part I of the act are governed by section 16(3)(b) and (d) of the Act.

TABLE OF CASES CITED IN VOLUME 3

A.C.L. v. Southern Ry.,
321 ICC 314 98

Acme Fast Freight v. Western Freight Association,
299 ICC 315 83

Adel Canning & Pickling Co. v. G. & F. R.R.,
287 ICC 239 286

Aetna Explosives Co. v. C. & E.I.R. Co.,
52 ICC 393, 396 291

Ahnapee & Western Ry. v. Akron & B.B. R.R.
302 ICC 265 245

Akers Motor Lines v. Lady Cornell Comb Co.,
203 F. Supp. 156 208

Albany Port District Comm. v. Ahnapee & W. Ry. Co.
219 ICC 151 .... 28, 34

Alcoholic Liquors, 6 Year Stop in Transit for Aging,
305 ICC 798 149

American Barge Line v. A.G.S. R.R.,
301 ICC 463 100

American Cast Iron Pipe Co. v. A.B. & C. R.R. Co.
201 ICC 454 62

American Cotton Waste & Linter Exchange v. B.&O. R.R. Co., et al.
169 ICC 710 262

American Home Foods Case
303 ICC 655 161

Ark. Farmer Plant Food Co. v. A.C.L. R.R.,
296 ICC 289 282

Arkansas Oak Flooring Co. et al. v. L. & A. Ry. Co.
334 U.S. 828 216

Armour & Co. v. A.T.&S.F.Ry.,
291 ICC 223 112

A.T.&S.F.Ry. v. U.S.&ICC
300 Fed. Supp. 1351 140

Bacon Brothers, et al. v. Alabama Great Southern R. Co., et al.
263 ICC 587 114
266 ICC 303 115
269 ICC 571 116

Baer Bros. Mercantile Co. v. Mo. Pac. Ry. Co. and D.&R.G.R.R. Co.
13 ICC 329 42

Baltimore & Ohio R.R. Co. v. U.S.
305 U.S. 507 176

Beaman Elevator Co. v. C. & N.W. Ry. Co.
155 ICC 313 43

Beloff Contract Carrier Operation
1 MCC 797 42

Blish Milling Co. v. Alton R. Co.
232 ICC 331 257

B.&M.R.R. v. U.S.
208 F. Supp. 661 64

Board of Trade of Kansas City, Mo., et al. v. A.T.&S.F.Ry. Co., et al.
69 ICC 185 96

Bogue Supply Co. v. Nevada Copper Belt Ry. Co.
96 ICC 434 271

Boney & Harper Milling Co. v. Atlantic Coast Line R.R. Co., et al.
28 ICC 383 106

Boyertown Burial Casket v. D.L. & W.R.R.,
305 ICC 333 286

Brainerd Fruit Co. v. C.G.W. Ry.,
163 ICC 585 83

Brown, W.P. & Sons Lbr. Co. v. Louisville & N.R. Co.
299 U.S. 393 49

Cairo Board of Trade v. C.C.C.&St. L.Ry.Co.
46 ICC 343 105

Calcium Carbonate Co. v. M.P.R.R.,
329 ICC 458 274

Canton R. Co. v. Ann Arbor R. Co.
163 ICC 263 57

Capital Grain & Feed Co. v. I.C.R.R. Co.
118 ICC 732 106

Caruso, G. & Co., v. St. Louis-S.F. Ry. Co.
156 ICC 429 288

Central Pennsylvania Coal Producers Assn. v. B.&O.R.R. Co.
196 ICC 203 26

Central Railroad of New Jersey v. United States,
257 U.S. 247 141

Central Vermont Ry. v. United States
182 Fed. Supp. 516 19

Chamber of Commerce of the State of N.Y. et al., v. N.Y.C. & H.R.R.R. Co., et al.
24 ICC 55 .... 19, 39

Chicago I. & L. Ry. Co. v. International Milling
33 Fed. (2d) 636 270
43 Fed. (2d) 93 270

Chicago & Wis. Points Proportional Rates
17 M.C.C. 573 .... 41, 106

C.&N.W. Ry. Co. v. Stein Co.
223 Fed. Rep. 716 208

City of Sheboygan v. C. & N.W. Ry. Co.
227 ICC 472 51

Class Rates Chicago, Ill. to Texas
311 ICC 660 139

Class Rate Investigation, 1939
262 ICC 447 .... 98, 222

Class Rates, Mountain Pacific Territory, Class Rates, Trans-Continental Rail, 1950
296 ICC 555 98

Coarse Grains for Feeding in WTL Territory
266 ICC 773 152

Coffee from Galveston, Texas, and other Gulf Ports
58 ICC 716 135

Cohen-Schwartz Rail & Steel Co. v. M.L.&T.R.R.S.S. Co., et al.
59 ICC 202 135

Commodities Over Tidewater Express Lines,
2 M.C.C. 356 84

Commodities—Pan-Atlantic S.S. Corp.,
309 ICC 587 85

Consolidated Classification Case
54 ICC 1 222
262 ICC 447 .... 222, 223

Consolidated Forwarding Company v. Southern Pacific Co., et al.
9 ICC 182 56

Consolidated Truck Lines v. Fess Wettmeyer
83 M.C.C. 673 18

Consumers Power Co. v. A.&S.,
286 ICC 291 120

Coomes and McGraw v. Chicago, Milwaukee & St. Paul Ry. Co. et al.
13 ICC 192 259

Dallas Cotton Exchange v. A.T. & S.F. Ry. Co.
163 ICC 57 26

Darling & Co. v. N.Y.C. & St. L.R. Co.
213 ICC 418 276

Davis v. Allen (S.C.)
117 S.E. 547 214

Davis, as Agent, etc. v. Portland Seed Co.
264 U.S. 403 114

Deisel Wemmer-Gilbert v. P.R.R.,
218 ICC 137 .... 285, 288

Delmar Case (See Great Northern Ry. Co. v. Delmar)

Diamond Mills v. Boston & Maine R.R. Co.
9 ICC 311 57

Dow Chemical Co. v. C.&O.Ry.,
306 ICC 403 161

D.S.S.&A. v. Mackinac Transp.
306 ICC 553 65

Duluth Chamber of Commerce v. C.B. & Q.R.R.,
210 ICC 652 286

Duplicate Payment of Freight Charges;
No. 36062 350 ICC 513 218

Dupont de Nemours Powder Co., E.I. v. Wabash R.R.Co.
33 ICC 507 63

Durez Plastics and Chemicals v. C.M. St.P.&P.R.R.,
259 ICC 335 275

Eastern Aircraft v. Olson & Son,
47 M.C.C. 363 287

Eastern Class Rate Investigation
164 ICC 314 27

Elevation Allowance at points located upon the Missouri, Mississippi and Ohio Rivers and on the Great Lakes
24 ICC 197 .... 154, 155

Empire Oil and Refining Co. v. A.T. &S.F.Ry.,
205 ICC 239 287
Equalization of Rates at North Atlantic Ports
311 ICC 689 28
Ex Parte 73
57 ICC 591 198
Ex Parte 104, Part VI, Warehousing and Storage of Property by Carriers at Port of New York, N.Y.
198 ICC 134 175
Ex Parte 235 292
Ex Parte 297
351 ICC 437 245
Export and Import Rates
169 ICC 13 31
Export and Import Rates to and from Southern Ports
205 ICC 511 31
Fancy Farms Case (See Bacon Brothers et al. v. Alabama Great Southern R. Co., et al.)
Farmers Union Grain Term. Ass'n. v. Can. Nat'l. Ry.,
315 ICC 559 279
Fernandez, E. & Co. v. S.P.R.R. Co.
104 ICC 193 18
Fifth Class Rates Between Boston, Mass. and Providence, R.I.
2 M.C.C. 530 139
Firestone Tire & Rubber Co. of California v. Southern Pacific Co.
243 ICC 157 271
Fraser-Smith Co. v. G.T.W. Ry. Co.
185 ICC 57 105
Fruen Grain Co. v. LaCrosse & S.E. Ry. Co.
132 ICC 747 271
Galveston Commercial Asso. v. G.H. & S.A.Ry. Co.
128 ICC 349 .... 31, 33
132 ICC 95 .... 31, 33
Gar Wood Industries v. A. & S.R.R.,
263 ICC 611 279
Glass Co., C.A. v. Pacific Electric Ry.,
256 ICC 541 120
Grain Elevation Allowances at Kansas City, Mo., and Other Points,
341 ICC 142 155
Grain & Grain Products
205 ICC 301 152
Great Northern Ry. Co. v. Delmar Co.
283 U.S. 686 283
Great Northern Ry. Co. v. Sullivan
294 U.S. 458 44
Great Western Packers Exp. Inc., v. U.S.
246 F. Supp. 15 48
Greater Des Moines Committee, Inc. v. Director General, as Agent, C.M. & St. P. Ry. Co., et al.
60 ICC 403 102
Guerdon Industries v. P.R.R.,
315 ICC 277 274
Gulf, Colorado & Santa Fe v. Hefley
158 U.S. 98 260
Hart v. Pennsylvania R.R. Co.
112 U.S. 331 160
Hausman Steel Co. v. Seaboard Freight Lines
32 M.C.C. 31 139
Healy & Towle v. C.&N.W. Ry. Co.
43 ICC 83 60
Hewitt-Robbins, Inc. v. Eastern Freightways
371 U.S. 84 .... 293, 294
High Point Chamber of Commerce v. Southern Ry. Co. et al.
314 ICC 683 45
Hocking Valley Railway Co. v. Lackawanna Coal & Lumber Co.
224 Fed. Rep. 930 .... 100, 260
Holt, Geo. C. and Odell, Benjamin B. as Receiver of Aetna Explosives Co. et al. v. C.&E.I.R. Co.
52 ICC 393 291
Hope Brick Works v. A.T.&S.F.Ry.
200 ICC 215 285
Hope Cotton Oil Co. v. T.&P. Ry. Co.
12 ICC 265 110
Hudson River Day Line Com. and Contr. Car. Applic.
250 ICC 396 42
Humphreys-Godwin Co. v. Y. & M.V. R. Co.,
31 ICC 25 113

Hurlburt, Frank, v. L.S. & M.S. Ry. Co.
2 ICC 81 261
Hyman-Michaels v. C.R.I.&P.R.R.,
308 ICC 339 276
ICC v. N.Y.N.H.&H. R.R.
372 U.S. 744 86
Illinois Central R.R. Co. v. W.L. Hoopes & Son, et al.
233 Fed. Rep. 135 208
Illinois Freight Assn. Agreement
283 ICC 17 245
Illinois Steel Co. v. Baltimore & Ohio R.R. Co.
320 U.S. 508 211
Increased Common Carrier Truck Rates in East
42 M.C.C. 633 84
Increased Rates, 1920
58 ICC 220 30
Indian Refining Co., Inc. v. L.&N. R.R. Co., et al.
112 ICC 732 100
Ingalls v. Maine Central R. Co.
24 Fed. (2d) 113 121
In Re Differential Rates
11 ICC 13 34
In Re Elevation Allowances
24 ICC 197 .... 154, 155
Interior Iowa Cities Case
28 ICC 64 103
Intermediate Rate Association v. Director General, Aberdeen & Rockfish R.R. Co., et al.
61 ICC 226 98
Interstate Commerce Commission v. Alabama, Midland Ry. Co., et al.
168 U.S. 144 131
In the Matter of Allowances to Elevators by the U.P.R.R.
12 ICC 85 154
In the Matter of Form and Contents of Rate Schedules and the Authority for Making and Filing Joint Tariffs,
6 ICC 267 56
In the Matter of Tariffs Containing Joint Rates and Through Routes for the Transportation of Property between Points in the United States and Points in Foreign Countries,
337 ICC 625 .... 7, 66, 67
In the Matter of Through Routes and Through Rates
12 ICC 163 .... 42, 43, 46, 96
Interchange between McLean Trucking and Manning Motor Express
340 ICC 38 46
Increased Common Carrier Truck Rates in East,
42 M.C.C. 633 82
Iron and Steel from Official to W.T.L. Territory
318 ICC 449 119
Jackson Common Carrier Application
19 M.C.C. 199 42
James Gallagher et al. v. P.R.R. Co.
160 ICC 513 175
Johns Manville Products Corp. v. Alton R.R.
269 ICC 739 287
Jubitz, G.L., Assignee, v. Southern Pacific Ry. Co. et al.
27 ICC 44 108
Kingan and Company v. Olson Transp. Co. et al.
32 M.C.C. 11 121
Koppers Co. v. C. & O. R.R.,
301 ICC 383 101
Larabee Flour Mills Corp., et al. v. A.T.&S.F.Ry. Co., et al.
148 ICC 5 96
Less than Truck Load Arbitraries from and to Minnesota Points,
309 ICC 527 82
Leonard Crosset & Riley, Inc. v. Akron C. & Y. Ry. Co.
176 ICC 309 115
Louisiana & A. Ry. v. Expert Drum Co.
228 F. Supp. 89 25
L. & N. R.R. Co. v. Central Iron & Coal Co.
265 U.S. 59 211
L. & N. Railroad Co. v. Maxwell
237 U.S. 94 260
L. & N. Railroad Company v. Mottley
219 U.S. 467 207
Lustberg Nast & Co., Inc., v. New York, N.H. & H. R. Co.
229 ICC 684 .... 271, 292
Lynchburg Traffic Bureau v. Smith's Transfer Corp.,
310 ICC 503 139

McCormick Warehouse Co. v. P.R.R. Co.
95 ICC 301 174
148 ICC 299 174
191 ICC 727 175

McLean Trucking v. U.S.
346 F. Supp. 349 46

Meat and Packing House Products to the South,
313 ICC 464 134

Merchants Warehouse Co. et al. v. U.S., ICC et al.,
283 U.S. 501 175

Meridian Rate Case
66 ICC 179 132

Mich. Buggy Co. v. G.R. & I. Ry. Co.
15 ICC 297 110

Middleton, Inc. v. N.S.R.R.,
215 ICC 411 287

Middlewest Freight Bureau v. C.B. & Q. R.R.,
309 ICC 303 274

Midwest Industries v. B. & O.,
303 ICC 319 286

Mid-Western Motor Frt. Tariff Bureau, Inc. v. Eichholz
4 M.C.C. 755 62

Milburn Wagon Co. v. L.S. & M.S. Ry. Co.
18 ICC 144 110

Miller & Lux, Inc. v. Southern Pacific Co.
102 ICC 137 271

Milwaukee Chamber of Commerce v. Flint & Pere Marquette R.R. Co., et al.
2 ICC 393 95

Miner Lbr. Co. v. Pennsylvania R. Co.
161 ICC 801 288

Missouri & Illinois Coal Co. v. Illinois Central R.R. Co.
22 ICC 39 39

Mixed Car Dealers Assn. v. D.L. & W. R.R. Co., et al.
33 ICC 133 262

Morgan v. Missouri, K. & T. Ry. Co.
12 ICC 525 121

Morris Buick Co., Inc. v. Grand Trunk W.R. Co.
229 ICC 109 .... 266, 270

Motor Carrier Rates in New England,
8 M.C.C. 287 84

Motor-Rail-Motor Traffic in East and Midwest
219 ICC 245 140

Movement of Highway Trailers by Rail,
293 ICC 93 63

Murphy, Andrew & Son, Inc., et al. v. Ann Arbor R.R. Co., et al.
147 ICC 499 272

Murray Co. of Texas v. Morrow, Inc.
54 M.C.C. 442 .... 284, 285

National-American Wholesale Lumber Assn. Inc. v. A.C.L. R.R. Co., et al.
120 ICC 665 60

National Furniture Traffic Conference v. Associated Truck Lines
332 ICC 802 45

Nelson, N.O. Mfg. Co. v. Mo. Pac.
153 ICC 272 262

New England S.S. Co. Com. Car. Applic.
250 ICC 184 42

New Jersey Central R.R. Co. v. MacCartney (N.J.)
52 Atl. 575 214

News Syndicate Company v. N.Y.C. R.R. Co., et al.
275 U.S. 179 21

Newton Gum Company v. C.B. & Q. R.R. Co., et al.
16 ICC 341 261

New York Central R. Co. v. Tal. Long Isl. R. Co.
288 U.S. 239 39

New York Central R. Co. v. Transamerican Petroleum Corp.
108 Fed (2d) 994 214

N.Y.N.H. & H. R.R. v. U.S.
199 F. Supp. 635 86

North American Cement Corp. v. Western Maryland Ry. Co.
129 ICC 90 121

Nueces County Nav. District v. A.T. & S.F.,
315 ICC 155 25

Official Southern Divisions
287 ICC 497 64

Oil Country Iron or Steel Pipe, Midwest to Oklahoma and Texas,
326 ICC 511 140

Old Colony Furniture Co. v. B. & M. R.R.
270 ICC 373 273
Ontario Paper Company, Ltd., et al. v. Canadian National Railways, et al.
95 ICC 66 22
102 ICC 365 22
Pacific Coast Fourth Section Applications
129 ICC 3 136
Parkersburg Rig & Reel Co. v. B. & O.R.R. Co.
225 ICC 581 55
Patterson, J.W., et al. v. L. & N. R.R. Co., et al.
269 U.S. 1 .... 118, 121
P.C.C. & St. L. Ry. Co. v. Alvin J. Fink
250 U.S. 577 .... 210, 212
People's Express Co.,
311 ICC 515 162
Pillsbury Flour Mills Co. v. Great Northern Ry. Co.
25 Fed (2d) 66 272
Pitts, H.B. & Sons v. St. L. & S.F. R.R. Co., et al.
10 ICC 684 262
Poor, A.J. Grain Co. v. C.B. & Q. Ry. Co., et al.
12 ICC 418 258
Porter Co., Inc. v. Central Vermont Ry.,
366 U.S. 272 19
Port of New York Authority v. Baltimore & O.R. Co.
248 ICC 165 34
Producers Cooperative Oil Mill v. St. L.S.F.,
322 ICC 267 282
Rags and Paper to Newark, N.Y.
208 ICC 327 136
Railroad Company v. Evans (Mo.)
228 S.W. 853 214
Rate Bureau Investigation—Ex Parte 297
351 ICC 437 245
Red Collar Line, Inc., Com. Car. Applic.
250 ICC 785 42
Refund Provisions, Lake Cargo Coal
220 ICC 659 99
Released Rates Order No. MC-607, 1965 158
Released Ratings on Engines,
287 ICC 419 161
Released Rates on Stone in the Southwest
93 ICC 90 162
Republic of France v. Director General, as Agent
81 ICC 174 273
Rogers v. Cincinnati, N.O. & T.P. Ry. Co.
258 ICC 553 266,271
Routing on Paper from Michigan to Illinois and Wisconsin Points
91 ICC 452 292
Routing Restrictions Over Seatrain Lines
296 ICC 767 40
Rush Common Carrier Application
17 M.C.C. 661 44
Salt Cake to Southern Territory
237 ICC 227 135
Sauers Milling Co. v. L. & N. R.R. Co., et al.
216 ICC 358 273
Scoular-Bishop Grain Co. v. M. & St. L. R.R. Co.
200 ICC 665 55
Service Pipe Line Co. v. C & N.W.
301 ICC 545 276
Seymour, John S. v. M.L. & T.R.R. & S.S. Co.
35 ICC 492 18
Sioux City Foundry and Boiler Co. v. C.B. & Q. R.R.,
318 ICC 13 282
Slane, O.W. Glass Co. v. Va. & S.W. Ry. Co.
39 ICC 586 113
Smith Mfg. Co. v. C.M. & G. Ry. Co.
16 ICC 447 110
Smoke Flue Cleaning Compounds, Transcontinental
268 ICC 619 274
South Atlantic Traffic Bureau v. Seaboard Air Line R.R.,
300 ICC 313 60
Southern Railway Company v. Southern Cotton Oil Co.
91 S.E. 876 210
Southern Roads Co. v. Galveston H. & S.A. Ry. Co.
168 ICC 768 57

South Texas Cotton Oil v. A. & S. R.R.
297 ICC 767 152
Southwest Fabricating and Welding Co. v. Alton & S.R.,
302 ICC 440 98
Spear Mills, Inc. v. Alton R. Co., et al.
268 ICC 330 276
Stevens Grocer Company, et al. v. St. L.I.M. & S. Ry. Co., et al.
42 ICC 396 .... 103, 104
Stoppage in Transit, Central Territory
51 M.C.C. 25 152
Sumter Packing Co. v. A.C.L. R.R. Co.
157 ICC 137 26
Swift and Company v. Alton R. Co.
263 ICC 245 263
258 ICC 103 263
Swift and Company v. P.R.R. Co., et al.
24 ICC 464 43
Terminal Warehouse Co. v. U.S.
31 Fed (2d) 951 175
Texas and Pacific Railway Co. v. Abilene Cotton Oil Co.
204 U.S. 426 259
Thomson v. U.S.
343 U.S. 549 37
T.I.M.E., Inc. v. U.S.,
359 U.S. 464 .... 292, 294
Tobin Packing Co. v. B. & O. R.R.,
299 ICC 221 280
Traffic Bureau, Merchants Exchange of St. Louis v. C.B. & Q. R.R.
14 ICC 317 155
Transcontinental Cases of 1922
74 ICC 48 133
Transcontinental Rates
46 ICC 236 136
Transit and Mixing Rules on Foodstuffs,
270 ICC 164 141
Transit Privileges at Ransom, W.Va.
63 M.C.C. 660 153
Tri-State Pipe Co. v. Abilene & Southern Ry. Co., et al.
234 ICC 55 262
United Shoe Machinery Corp. v. Dir. Gen. and Pennsylvania R.R. Co.
55 ICC 253 272
United States v. G.N.R.R.,
337 Fed (2d) 243 274
United States v. Joint Traffic Assn.
76 Fed. Rep. 895 239
171 U.S. 505 239
U.S. v. N.P. Ry.
301 ICC 581 261
United States v. Reading Co.
226 U.S. 324 239
United States v. Southern Ry. Co.
142 F. Supp. 741 289
United States v. Trans-Missouri Frt. Assn., 17 Sup. Ct. Rep. 540
166 U.S. 290 239
United States v. Union Pac. R. Co.
226 U.S. 61 239
Upjohn Co. v. Pennsylvania R.R.
306 ICC 325 161
U.S. Industrial Alcohol v. Director General,
68 ICC 389 275
Van Dusen Harrington Co. v. Northern Pac. Ry. Co.
32 Fed. (2d) 466 272
Vegetables from Florida to Eastern Points
219 ICC 206 114
Vermont Transit Co., Inc.
11 M.C.C. 307 44
Victory Granite Co. v. Central Truck Lines, Inc.
44 M.C.C. 320 121
Von Platen Box Co. v. Ann Arbor R. Co.
214 ICC 432 113
Waverly Oil Works Co. v. P. R.R. Co., et al.
28 ICC 621 50
West Virginia Rail Co. v. I.C. Ry. Co.
53 ICC 21 135
Westinghouse Electric Corp. v. C.G.W. Ry.,
299 ICC 527 287
West Petroleum Co., A.E. v. A.T. & S.F. Ry.,
212 Fed. 2nd 812 289
Weyl-Zuckerman & Co. v. Alabama, T. & N.R. Corp.
210 ICC 565 91
Windsor Turned Goods Co. v. C. & O. Ry. Co.
18 ICC 162 110

INDEX

Overcharges

Payment of Charges